MARS

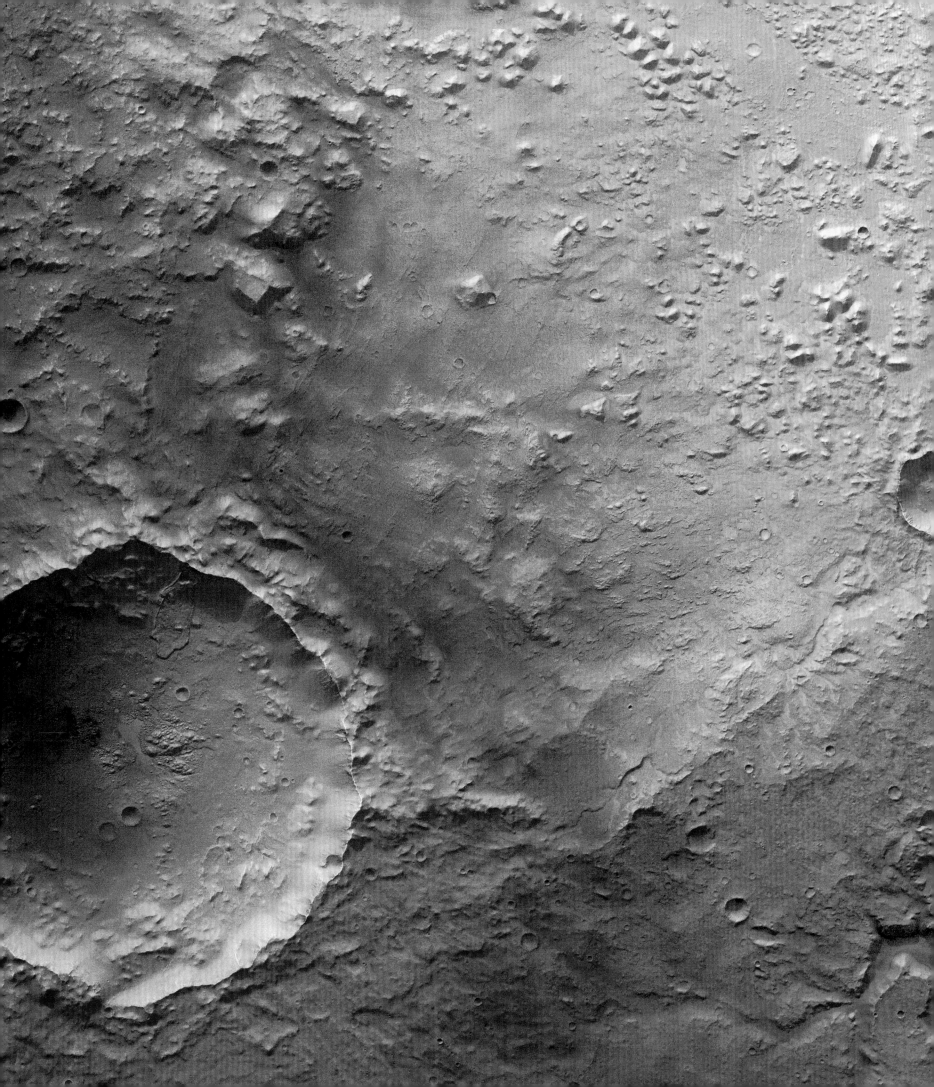

MARS

THE ULTIMATE GUIDE TO THE RED PLANET

Giles Sparrow

Quercus

Contents

QuercusEye

This book is enhanced with exciting new technology, allowing you to unlock moving video footage from images in the book using most web-enabled smartphones or tablets. Simply download and open the free QuercusEye app and locate the pictures with the QuercusEye icon. These can be found on pages 26, 109, 187, 202, and 210. Hover the camera above each image so the picture fits the screen, and watch it come to life.

Go to www.quercuseye.com for more details.

Introduction

In 1609 German philosopher, mathematician, and mystic Johannes Kepler published a book that began a revolution, the *Astronomia Nova* or *New Astronomy*. Based on a brilliant analysis of meticulous observations by his former tutor, the Danish astronomer Tycho Brahe, Kepler showed that planets move not on circular paths but along ellipses, following three simple laws. In the process, he laid the foundations for Isaac Newton's laws of motion and gravitation, and the age of Enlightenment that began in the late 17th century.

But the full title of Kepler's work reveals his most important inspiration —*A New Astronomy based upon causes – or Celestial Physics, revealed by commentaries on the motions of the Star Mars by the Noble Tycho Brahe*. The largest word on the title page is "Martis"—the Latin word for Mars.

Why Mars? How has the Red Planet played such an important role in our growing understanding of our place in the Universe, why does it haunt our imaginations so much, and what role could it play in the very future of humanity itself? For Brahe, Kepler, and the other pretelescopic astronomers, the fascination of Mars lay in the planet's uniquely erratic dance through the sky—speeding up in some places, slowing down in others, and sometimes reversing its motion completely. This, as Kepler discerned, arises because of its place in the solar system and its notably elliptical orbit.

But for later astronomers, the Red Planet's allure lay in its proximity and tantalizing similarity to Earth. To our modern eyes, spoilt by astonishing space-probe images such as those that grace the pages of this book, these general similarities may seem to dissolve into the myriad stark differences of a beautiful yet alien world, but step back, and the sibling resemblance becomes more obvious. First and foremost, Mars is a rocky world like the Earth—a solid body that despite its smaller size and mass still has enough gravity to hold onto an atmosphere that shows complex, changing weather patterns. It is neither a bloated ball of gases like the outer worlds of the solar system, nor a charred and barren rock like the innermost planet Mercury. What's more, at its closest to the Sun, Mars lies just 35 percent further from the Sun than our own planet—so that while its surface temperature is naturally colder, it is still far more hospitable than the searing heat and choking atmosphere of Earth's inner neighbor, Venus.

As the only other rocky planet on which we can see surface features, it seems as though the secrets of Mars lie enticingly within our reach. Little wonder, then, that early telescopic observers strained at the eyepiece to sketch vague patterns of dark and light on the Martian disk. Most concluded early on that Mars was a desert world, but they also recognized the changing white spots at its north and south poles as ice caps similar to those of Earth's Arctic and Antarctic regions, and correctly realized that the Red Planet was not entirely dry.

"Mars alone enables us to penetrate the secrets of astronomy which otherwise would remain forever hidden from us." JOHANNES KEPLER

So when they understandably reached the conclusion that the dark areas, which seemed to wax and wane in size and intensity from month to month, might be great lakes, oceans, or even forests of vegetation, it's only the arrogance of hindsight that allows us to mock those who reported seeing artificial channels connecting these wetter areas, carrying water across the sandy wastes. Even today, the place of Mars in the popular imagination owes a huge debt to *The War of the Worlds*, H.G. Wells's 1897 tale of mankind brought low by alien invasion—and Wells in turn was directly inspired by contemporary scientific depictions of a dying world that have eerie echoes in modern discoveries.

Even in the 20th century, as evidence mounted that Mars was a cold and arid world with a tenuous atmosphere, it remained an object of fantasy and fascination for the first generation of rocket scientists. If there were no native Martians, then surely it was mankind's manifest destiny to *become* Martians? The dream of a colony on the Red Planet, far more than the barren and challenging wastes of our planet's own satellite, drove the engineers and designers who ultimately put men on the Moon. And while political and fiscal realities ultimately derailed their grand schemes, the robot probes that have explored Mars in our stead would not have been possible without the work of those early dreamers.

So, welcome to *Mars*—a unique book that hopefully does justice to a unique subject, showcasing the stunning imagery returned by more than a dozen groundbreaking space missions and putting it into historical, thematic, and geographical context. However, I hope that this volume provides not just a feast for the eyes, but also a digestible summary of the current state of research—the evidence, theories, and unanswered questions that make up our present portrait of the intriguing Red Planet. That picture is still continually evolving, and each question answered raises another in its place, showing why, four centuries on from Kepler's planetary breakthroughs, Mars is still a planet with many secrets to reveal, and a world that exerts a powerful hold on our imaginations.

Giles Sparrow

1 The fascination of Mars

The Red Planet has exerted a powerful fascination for stargazers since before the dawn of scientific astronomy. Viewed as a harbinger of war and ill omen, puzzled over for its complex dance across the heavens and ultimately studied as a world equal to our own and a possible home to other civilizations, Mars is a world of mysteries with many secrets yet to be revealed.

➔ The long scar of the Valles Marineris canyons dominates this distant view of Mars.

The prehistory of Mars

Mars is an ever-present feature of Earth's skies—a brilliant red "star" that was always considered to be worthy of note. Even before the invention of the telescope, the obsession with its curious behavior would transform our understanding of the Universe.

As one of the five planets visible to the naked eye, Mars was considered to be a special and important object from the time of the very first prehistoric stargazers. The word "planet" comes from the Greek for "wanderer," and refers to the behavior of these lights in the sky, which slowly moved against the "fixed" patterns of constellations that we now know to be made up of more distant stars. Specifically, the planets moved through the band of 12 constellations known as the zodiac. From the earliest times, people assigned attributes to the planets, and felt that their position in the zodiac could act as a mirror to events on Earth—the basis of astrology. Once this idea had taken hold, the ability to predict the motion of the planets became an obvious concern, since it might also allow events on Earth to be foretold.

The earliest astrologers in ancient Mesopotamia (modern Iraq) made a natural connection between Mars's red color and ideas of war and bloodshed, associating the planet with Nergil, the god of war and pestilence. The ancient Greeks, who inherited much of their astronomy from Mesopotamia, therefore associated the planet with Ares, and today we know it by the name of Ares's Roman counterpart.

However, Mars presented unique challenges to attempts at understanding its motion. For one thing, it made large "retrograde" loops, traveling backward in the sky for several months at a time before resuming its general eastward progression —an effect we now know to be caused when Earth "undertakes" its slower-moving outer neighbor. The planet also changed its speed more or less at random (thanks to its markedly eccentric orbit).

A more general problem for early attempts to understand *any* planetary motion was the widespread assumption that Earth was fixed at the center of the Universe, with the Sun, Moon, and planets orbiting on spherical shells. In order to explain the movements of Mars and other planets in this geocentric system, Greek philosophers had to introduce a system of "epicycles"—smaller

c.500 BC
Babylonian and Greek astrologers associated the Red Planet with their gods of war, Nergal and Ares, but it was the Romans who elevated Mars to the status of one of their major gods, and attached similar importance to "his" planet.

c. AD 1200
Medieval astrology associated the movements of the planets through the zodiac with various attributes, as depicted in this colorful woodcut showing the personifications of Mars, the Sun, and Venus.

suborbits, carried on the major spheres, around which the planets actually moved. This idea was brought to perfection by the Greek-Egyptian mathematician Ptolemy in the second century AD, and became an almost unchallenged article of philosophical and religious dogma for almost 1,500 years. But in the late medieval period, as improving technology allowed stargazers to make increasingly accurate observations of the path of the planets in general, and Mars in particular, it became apparent that no geocentric model could adequately describe these complex movements.

This led in 1543 to Polish cleric and astronomer Nicolaus Copernicus's proposal for a "heliocentric" Universe, with the Sun at its center and Earth as one of six planets orbiting around it. Yet, although this would seem to be a huge leap forward in understanding, the idea was slow to catch on, not only because of opposition from an entrenched religious and academic establishment, but also because, relying on perfectly circular orbits, it still required epicycles to explain Mars's varying speed.

In the late 16th century, however, the great Danish astronomer Tycho Brahe set out to chart the movements of Mars in more detail than ever before. This led him to develop a convoluted hybrid model of the Universe. After his death in 1601, it inspired his student, German mathematician and astrologer Johannes Kepler, to finally break free of the limits imposed by circular motion, proposing in 1609 that the orbits of Mars and the other planets were in fact ellipses, with the Sun at one of two "focus points" and the planets moving faster or slower depending on their distance from it.

Careful studies of Martian motion had led the way to a revolution in astronomy. Within a year, the arrival of the astronomical telescope would allow the first studies of the Red Planet itself.

DE MOTIB. STELLÆ MARTIS

1585
Although the idea of a
Sun-centered Universe began

1608
Kepler's intricate diagram
shows how the orbits of

Mars through the telescope

The invention of the first telescopes in the early 17th century revolutionized astronomy, and led to the realization that the other planets were worlds in their own right—and Mars was perhaps the most similar of all to our own planet Earth.

Thanks to its relatively small size, Mars always appears fairly tiny in our skies despite its occasional proximity to Earth—even at its closest and largest, it is still just $\frac{1}{60}$ the apparent diameter of the Moon. As a result, when Italian astronomer Galileo Galilei turned a telescope toward Mars for the first time in 1610, he still saw little more than a small, ruddy disk.

Galileo set himself the goal of observing Martian "phases": just as the Moon, Mercury, and Venus change their phases as we see varying amounts of their sunlit hemispheres, so it should be possible to peer "around" Mars at certain points in its orbit, revealing a small amount of its dark side and reducing the illuminated disk to a "gibbous" phase, similar to that shown by the Moon when it is not quite full. In reality, even this observation was far beyond the reach of Galileo's instrument, and it was not until 1645 that the Polish astronomer Johannes Hevelius successfully spotted Martian phases.

By this time, telescopes had grown in size and quality to the point where they could show the planet as a distinct disk (at least around its biennial oppositions, when Mars is closest to Earth and largest in the sky), and astronomers soon began to report features on the surface. Italian Jesuit priest Daniello Bartoli reported two dark patches on the surface in 1644, and two more Italians, Giovanni Battista Riccioli and Franceso Maria Grimaldi, reported a number of features over the following decades.

By 1659, Dutch astronomer and physicist Christiaan Huygens was able to produce the first map of Mars, and a year later he used the movement of Martian features to make the first measurement of the planet's rotation period. Another Italian, Giovanni Domenico Cassini, refined Huygens's figure in 1666, coming up with a Martian "day" just three minutes adrift of the currently accepted value of 24 hours 39 minutes. Cassini was the first to identify the

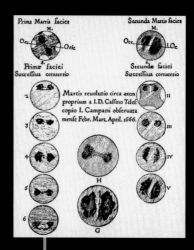

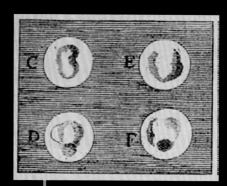

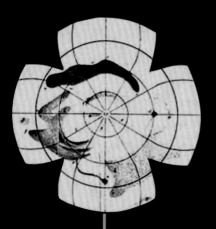

1666
Giovanni Domenico Cassini used these sketches of Mars to

1666
British scientist Robert Hooke drew his sketches of

1831
German astronomers Beer and Mädler transferred their

bright spot of an icy polar cap at the Martian south pole, and in 1672 Huygens found a similar bright patch at the north pole.

Telescopes would remain too small and limited for detailed observation for another century or more, but some ingenious observations were still possible, particularly involving the prominent Martian poles. In 1704, the Italian Jacques Philippe Maraldi reported brightness variations in the south pole through the course of a Martian day, indicating that the south polar cap was not entirely symmetrical, while in the 1780s, German-born British astronomer William Herschel, fresh from his discovery of the planet Uranus, showed that the southern cap varied in brightness from year to year. He correctly concluded that this was due to an ice cap varying its extent through the cycle of Martian seasons.

Breakthroughs in the design of optical instruments during the early 19th century finally brought Martian surface detail within view, and in 1830 two German astronomers, Wilhelm Beer and Johann Henrich von Mädler, set out to make a new map of the planet. They identified many features for the first time, and showed that they were permanent markings rather than shifting patterns.

Throughout the 19th century, astronomers continued to produce ever more detailed maps of the dark and light features on Mars (features we now know to be large-scale differences in surface brightness caused by the exposure of dark basaltic rock in some places but not others).

At the time, seasonal changes in the intensity and distribution of the dark features (caused in reality by shifting dust and changing illumination) were widely taken as evidence that they were related to vegetation. The first Martian weather feature (named the Blue Scorpion), was spotted by Italian astronomer Fr. Angelo Secchi in 1858, and at this time, the true nature of the thin, dry atmosphere was not understood.

A decade later, Secchi also became the first to use the term *canali* in relation to the planet's surface features, describing dark linear features that he had observed. However, it was not until 1877 that the infamous "canals of Mars" burst upon the world, thanks largely to the work of another brilliant Italian observer, Giovanni Schiaparelli (see over). That same year, US astronomer Asaph Hall discovered not one, but two small satellites of Mars, soon named Phobos and Deimos (fear and terror) in honor of the mythological war god's sons.

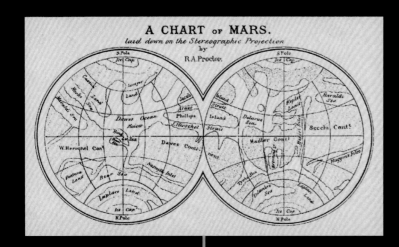

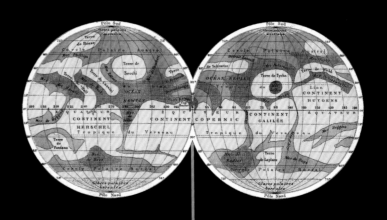

1867
Richard Proctor's map was based on drawings by another British

1881
The favorable oppositions of 1877 and 1881 led to the production

The coming of the Martians

A series of genuine discoveries and mistaken observations during the late 19th century saw the birth of an obsession with intelligent life on Mars—a dream that would survive until well into the 20th century.

In 1877, Mars came closer to Earth than it had done for many years, and Italian astronomer Giovanni Schiaparelli was one of many who set out to map the planet in new detail. In between the various dark features he charted, he reported dozens of linear connecting features—*canali*—that he named after our own planet's famous rivers.

Unfortunately, while the Italian word *canali* means simply a trough or channel for water, its most obvious English equivalent carries with it widely accepted connotations of artificial construction. Nevertheless, the word "canals," though used in the first translations of Schiaparelli's work, did not spark an immediate assumption that Schiaparelli was demonstrating evidence of intelligent Martians. As he put it: "It is not necessary to assume that [the canal network] was the work of intelligent beings, and in spite of the geometric appearance of the whole system, we are inclined to believe that it originated during the evolution of the planet, just as on Earth the English Channel or the Channel of Mozambique."

The idea that broad waterways of some sort connected the polar ice caps with dark areas of supposed vegetation was not in itself so outrageous, and had been proposed even before Schiaparelli's observations, notably by French astronomer Camille Flammarion in a work of 1876 that drew on hundreds of previous maps and drawings. Flammarion's friend Henri de Parville had already published perhaps the first fictional account of a Martian civilization in a book of 1865, and Flammarion himself was not averse to dabbling in fiction. However, he was also a widely read author of popular works about astronomy, and did much to promote the serious consideration of intelligent Martians in works such as *The planet Mars and Its Conditions of Habitability* (1892).

Despite this, in the years following the publication of Schiaparelli's map, there was growing doubt about the reality of the canals. For most of the skeptics, the problem was not so much a principled objection to the idea of life on other planets, as their inability to replicate Schiaparelli's

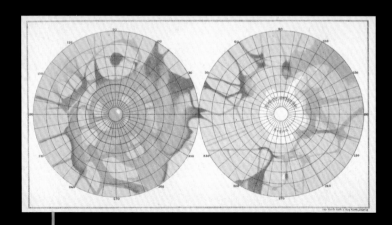

1881
Schiaparelli's detailed map of the two Martian hemispheres,

1897
The publication of H.G. Wells's *The War of the Worlds* introduced the

observations. The astronomical community soon divided into two camps, with some experienced observers reporting ever more elaborate canal networks linking to artificial seas at equatorial latitudes, while others (including the noted American E.E. Barnard, famed for his sharp sight) insisted they saw no straight lines on Mars at all. It was not until 1903 that the controversy was settled after astronomers J.E. Evans and E.W. Maunder of the Royal Greenwich Observatory devised an ingenious experiment using pupils from a nearby school to show that some people have a natural tendency to imagine straight lines joining isolated dark patches on a surface such as the disk of Mars. In other words, the canals are an optical illusion.

Nevertheless, by this time the idea of Martian canals and intelligent Martian life had well and truly taken hold. Some astronomers, most notably Percival Lowell, founder of the observatory at Flagstaff, Arizona, refused to accept that the canals were illusory, producing ever more elaborate maps and descriptions of the likely nature of Martian civilization.

Others, meanwhile, allowed the prospect of Martian life to fuel their imagination. Enlightenment writers such as Athanasius Kircher and Emanuel Swedenborg had occasionally treated the prospect of contact with a Martian civilization as an intriguing philosophical problem or a subject for satire, and de Parville's story (which recounts the discovery of an alien "mummy" entombed within a meteorite) is perhaps the first work of "Martian fiction." However, it was H.G. Wells's famous novel of 1897, *The War of the Worlds*, that put the possibility of intelligent, and possibly hostile, creatures from Mars at the forefront of the popular imagination. While Wells did not mention the canals directly, he nevertheless painted a picture of the Red Planet that would have been familiar to readers of Flammarion and Lowell—a world dying from a long, slow drought, whose inhabitants were willing to go to extraordinary lengths to survive.

Wells's work was enormously influential, and has been frequently revisited and reimagined ever since. Others soon followed in his footsteps, telling stories of alien invasion and infiltration of Earth, or of humans forced to fend for themselves amid the strange environment and curious customs of Mars itself. Perhaps the most notable of Wells's successors was Edgar Rice Burroughs (also inventor of Tarzan), who wrote no fewer than 11 books in his "Barsoom" series between 1917 and his death in 1950, recording the adventures of John Carter, a veteran of the American Civil War who finds himself transported to Mars and caught up in adventure and intrigue.

Nearly all these early works of fiction accepted the general view of Mars as a harsh world that was nevertheless relatively Earthlike and capable of sustaining advanced life. But by the time Burroughs was writing his last Barsoom books, new discoveries had changed that picture beyond all recognition.

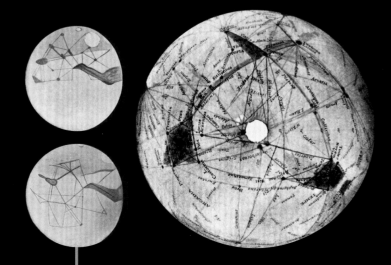

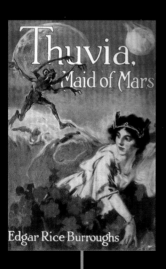

1905
Initial rough sketches by Percival
Lowell evolved into detailed charts

1917
Edgar Rice Burroughs's pulp
Martian adventure stories kept

Into the Space Age

The 20th century saw some remarkable seesawing in our understanding of Mars, much of it triggered by the rise of space technology and our newfound ability to study the Red Planet up close for extended periods of time, and even visit its surface.

The first hint that Mars might not be the potentially habitable world assumed by late 19th-century astronomers came as early as 1894, when American astronomer William W. Campbell used improved spectroscopic techniques to analyze the sunlight reflected off Mars and concluded that, contrary to the suggestions of earlier measurements, Mars did not have a substantial, water-laden atmosphere. Published at the height of interest in the Martian canals, Campbell's measurements were considered controversial and were widely ignored until repeated in 1925 by Walter S. Adams, another spectroscopy expert.

In fact, the idea of a relatively hospitable Mars suffered a number of blows in the 1920s. A year before Adams's measurement, astronomers from Mount Wilson Observatory were able to measure the planet's temperature for the first time, revealing a midday temperature range that varied between −90.4°F (−68°C) at the poles and 44.6°F

(+7°C) at the equator. In 1926, Adams directly confirmed that the levels of both oxygen and water vapor were very low, rendering the Martian surface a hostile desert, and in 1929, French astronomer Bernard Lyot confirmed—from the atmosphere's effects on sunlight—that it could be no more than 15 percent of Earth's atmosphere. (William Herschel had first concluded as early as 1783 that the Martian atmosphere must be thin after tracking the brightness of stars passing behind the planet.)

Another significant development in 1929 came with the publication of a new map of Mars compiled by Greek astronomer E.M. Antoniadi. This atlas drew on a century's worth of earlier maps and charts to inspire its naming conventions, and the names Antoniadi assigned became the de facto standard, finally given official sanction by the International Astronomical Union in 1960. Antoniadi was also the first person to suggest that the periodic yellow cloud features seen on Mars might be dust storms.

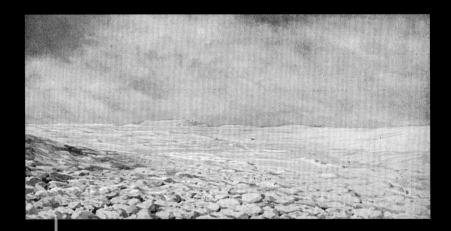

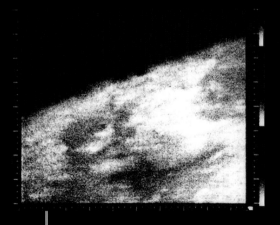

1930
By the 1930s, most astronomers had reluctantly concluded that Mars was an arid world with a desert landscape and sparse, cold atmosphere, as depicted in this artist's impression from the time.

1965
Mariner 4 sent back the first close-up images of Mars from a distance of less than 6,200 miles (10,000 kilometers). As the probe passed over the southern highlands, they revealed a Moonlike, cratered terrain.

gas (commonly known as dry ice) could not be a major constituent of the polar ice caps—instead, they must be pure water. This error would not be corrected until 1967, when the Mariner 7 space probe flew past the planet and measured the spectrum of the polar ice for the first time (for more on the complex structure of the polar caps, see page 40).

Since the dawn of the Space Age in 1957, our understanding of Mars has been hugely enriched by data from a variety of spacecraft. From the first brief flybys by space probes in the 1960s, they have evolved into sophisticated orbiters, landers, and rovers, whose exploits are chronicled in detail in Chapter 4.

Even so, ground-based, or at least Earth-based, studies can still make a substantial contribution. Since the 1980s, scientists have come to realize

particularly useful for monitoring global weather patterns and observing how dust storms develop.

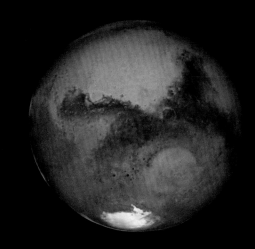

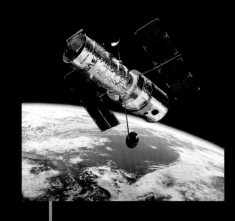

1976

The twin Viking Mars landers sent back the first images from the planet's surface in late 1976, revealing rust-colored, rock-strewn landscapes, and testing the soil for signs of life.

1990

The launch of the Hubble Space Telescope allowed astronomers to study Mars in considerable detail from Earth's orbit, especially around the time of its oppositions (close approaches to Earth).

2012

The successful touchdown of NASA's Mars Science Laboratory, better known as the Curiosity rover, in Gale crater marks the apogee of Martian exploration … so far.

2 World of wonders

Mars is an intricate planet, shaped by a wide variety of different forces that create a delicate interplay of geology and climate to rival Earth's own. Its surface shows the influence of many of the same factors that also shape our home world—tectonics, volcanism, and glaciation, and the actions of wind, weather, and seasons. The result is a world of awe-inspiring beauty and wondrous complexity.

→ Sunrise over Gale Crater, landing site of NASA's Curiosity rover.

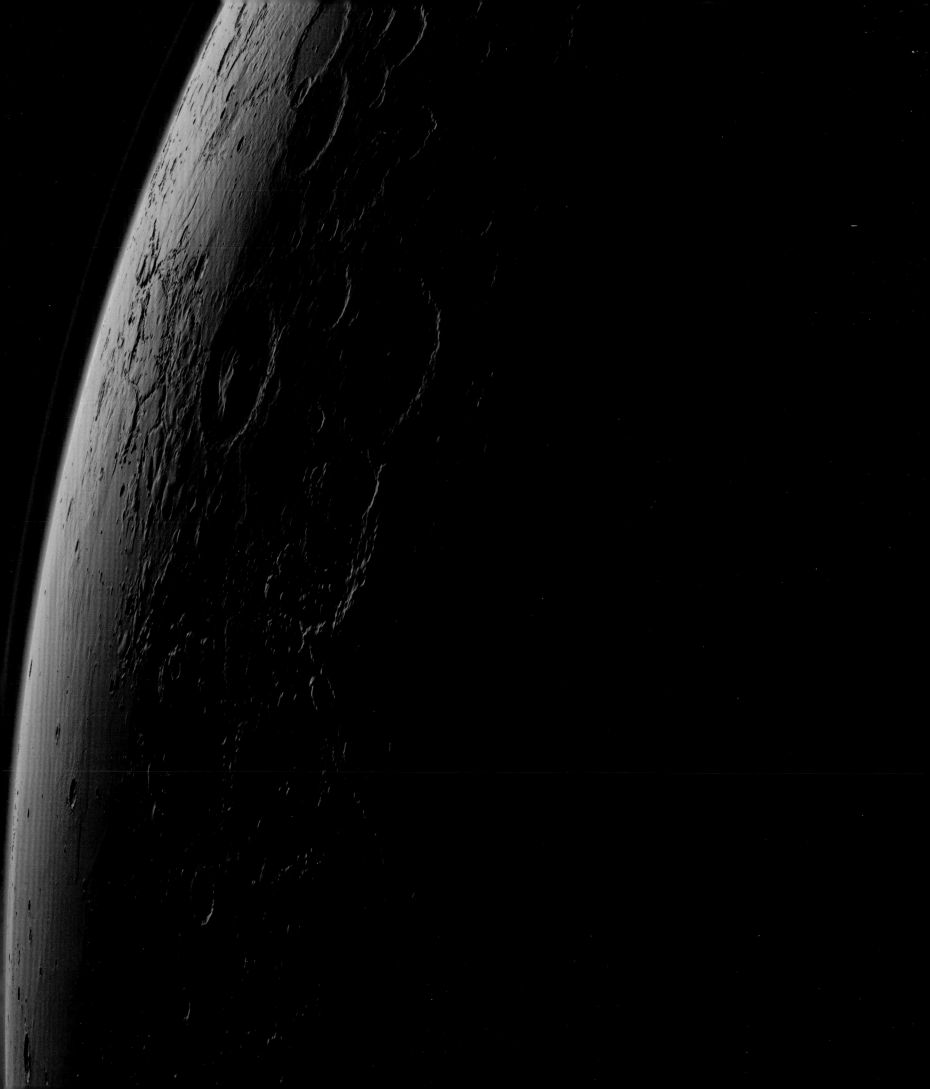

Mars in the solar system

Mars is the fourth planet from the Sun, outermost of the solar system's terrestrial worlds. Physically, it is the second smallest of these rocky planets, with a diameter of 4,220 miles/6,792 km (just over half that of Earth), a mass 11 percent of our planet's, and surface gravity a mere 38 percent of Earth's.

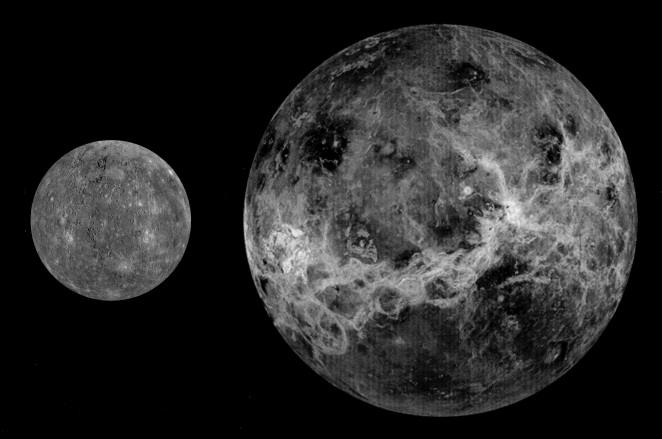

The Red Planet's orbit takes it around the Sun once every 687 days (1.88 Earth years), with an average distance from the Sun of 142 million miles (228 million kilometers). Unlike Earth's orbit, however, Mars's path through space is notably elongated or elliptical, ranging between (128.4 million miles (206.7 million kilometers) from the Sun at perihelion (closest approach) and 154.8 million miles (249.2 million kilometers) at aphelion (greatest distance). Astronomers often refer to interplanetary distances in terms of astronomical units (AU), where 1 AU is the average distance between Earth and Sun, equivalent to 93 million miles (150 million kilometers). In these terms, Martian perihelion is at 1.38 AU, while aphelion is at 1.67 AU.

As Mars and Earth orbit around the Sun, the distance between them is constantly changing: when the two planets are lined up on the same side of the Sun, and Mars lies opposite the Sun in the sky—an event called "opposition,"—our two worlds are closest together, but thanks to the markedly elliptical Martian orbit, and the slight eccentricity of our own orbit, the distance of this close approach can vary hugely, between 0.36–0.69 AU (33.5–64 million miles or 54–103 million kilometers). When Mars and Earth are on opposite sides of the Sun, a so-called "superior conjunction," the distance between them is at its greatest—anything between 1.36–1.69 AU (127–157 million miles or 204–253 million kilometers).

This complex relationship between orbits also causes Mars to vary substantially in brightness and apparent size. At its most brilliant, Mars can reach −3 on the scale of astronomical magnitudes, a scale in which lower numbers are brighter than higher ones. At such times, it outshines every star in the sky and is the second brightest planet after Venus. At its faintest, it can fade to magnitude +1.6, leaving it outshone by the two dozen brightest stars. The planet's angular diameter, which is imperceptible to the naked eye but important for telescopic observations, ranges between 25.1 seconds of arc at the closest oppositions and 3.5 seconds of arc at the most distant superior conjunctions (a second of

arc is $\frac{1}{3600}$ of a degree, or roughly $\frac{1}{1800}$ the diameter of the Sun or Full Moon). More details of Mars's motion through the sky, and some tips for practical observing, are given in the appendix on page 218.

The Martian day is known as a "sol" and is very similar to Earth's, at 24 hours and 37 minutes, while its axis is tilted at 25.2 degrees from "upright" —close to Earth's own figure of 23.4 degrees. The planet's axial tilt alone would be enough to produce an Earthlike pattern of seasons, but its varying distance from the Sun adds to its seasonal complexity (see page 64), as do slow changes in both the tilt and the shape of the Martian orbit. The overall result is a series of interlocking periodic variations known as Milankovitch Cycles (see page 66).

On balance, however, Mars's greater distance from the Sun puts it at the outer edge of our solar system's "habitable zone"—the region where liquid water can exist on planetary surfaces. The main reason for today's lack of surface water is probably

the thin Martian atmosphere, which causes any liquid exposed to the air to evaporate rapidly (see page 62).

Orbiting at the edge of what astronomers usually call the inner solar system, Mars defines the inside edge of the broad zone known as the asteroid belt. The next planet out, Jupiter, orbits at between 4.95 and 5.45 AU from the Sun, and the gap is filled by countless small worlds, most just a few miles across if not smaller. In general, these worlds keep their distance from both Mars and Jupiter, but occasional close encounters with each other or with the gravity of either planet can kick them into orbits that cross that of Mars and may even collide with the planet. Mars's two small, irregularly shaped satellites, Phobos and Deimos, were once thought to be stray asteroids caught up by Martian gravity during a close encounter with the Red Planet, but as we shall see (page 76), their true history may be far more interesting.

↑ Mars is the fourth and second smallest of the rocky planets, worlds that share common features such as a solid crust and a layered interior, but show stark differences in their surface conditions. Mercury (far left) is an airless ball of baking rock, Venus a hellish world boiling under a dense acidic atmosphere, and our homeworld Earth a temperate planet with plentiful liquid water (and its moon). Mars, though small, cold, and dry, is the most Earthlike of all the solar system's other worlds.

Birth of a planet

In the aftermath of the Sun's formation around 4.6 billion years ago, the newborn star was surrounded by remnants of the material out of which it had formed. Flattened out into a broad, spinning disk, this debris formed a "protoplanetary nebula"—the raw material from which all the planets ultimately formed.

The density and composition of the disk varied with distance from the Sun, giving rise to the broad differences between planets that we still see in the solar system today. Close to the Sun, temperatures were high enough to cause volatile materials with low melting points (broadly known as ices, but not exclusively limited to water ice) to evaporate into gas. Under the weak but persistent pressure of solar radiation, these gases were driven out of the inner solar system, leaving behind only "refractory" dust with higher melting points. Further out, beyond the solar system's "frost line," temperatures were cold enough for ice and raw protoplanetary gas to persist, and since this vastly outweighed the dust, it became the dominant component of the disk.

Over millions of years, electrostatic forces among the dust particles close to the Sun caused them to draw together, gradually growing into small asteroids. Once a few of these objects grew large enough to develop their own substantial gravity, they began to pull in material from their surroundings and grow at a much faster pace, developing into "planetesimals" hundreds or thousands of miles across. These objects were the building blocks of Mars and the other rocky planets we know today—they rapidly soaked up the nebula's remaining material, and collided and merged to form true "protoplanets." Beyond the frost line, meanwhile, the giant planets either grew in more or less the same way—albeit with much

↓ A combined view from the Spitzer Space Telescope and the Atacama Large Millimeter Array captures a nearby newborn solar system, around the bright star Fomalhaut, at both visible (blue) and radio (orange) wavelengths. Fomalhaut's planetary system is at an early stage in its formation, but a large young planet is thought to be responsible for sweeping out debris from its orbit and creating the well-defined ring structure.

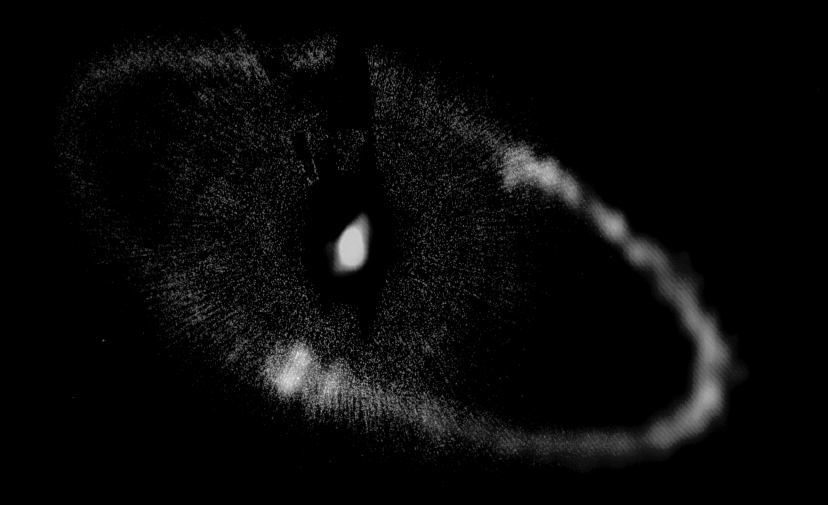

← This artist's impression depicts the early days of our solar system. As a cloud of protostellar matter collapsed under its own gravity, the central regions grew hotter and denser until eventually they became a fully-fledged "protostar." Collisions between clouds of gas in random motion gradually transformed the surrounding material into a flattened rotating disk, while the conservation of angular momentum caused the surrounding disk to spin more and more rapidly. Excess material falling onto the infant Sun was flung out from above and below the disk as a pair of "bipolar jets," while within the main disk, the nuclei of planets began to coalesce.

larger layers of gas and ice coalescing around solid protoplanetary cores—or formed from swirling clumps of protoplanetary material separating from the main nebula surroundings and collapsing from the "outside in."

Until quite recently, astronomers believed that the planets formed in more or less the same locations where they orbit today. The general trends in the size of planets were explained as variations in the thickness of the disk and the amount of material available at different distances. The relatively small size of Mars presented a problem, but along with the large gap between Mars and Jupiter occupied only by small asteroids, this was explained as an effect of Jupiter's formation disrupting the growth of Mars.

However, in the past decade, new research has suggested that the solar system's early history was actually far more dynamic, with Mars close to the center of the action. According to the so-called "Nice Model," first published in 2004, the solar system's four giant planets formed consid-erably closer to the Sun than they are now. Jupiter originated in the middle of the present-day asteroid belt and then spiraled inward, absorbing material

as it did and choking off the growth of Mars in the process. Thereafter, the giant planets spiraled outward again, until eventually (around 4 billion years ago), Uranus and Neptune began to disrupt the Kuiper Belt, a huge cloud of comets and small icy worlds that had formed at the outer edge of the solar system. Many Kuiper Belt Objects (KBOs) were flung out of the solar system altogether, while others were sent plunging toward the inner planets, where they rained down on the surfaces of these newborn worlds in an event known as the Late Heavy Bombardment (see page 26). By around 3.8 billion years ago, the orbits of the planets had more or less stabilized at last, and the Late Heavy Bombardment petered out, leaving the Kuiper Belt greatly depleted.

As we shall see, these cataclysmic events had important consequences for the development of the Martian landscape and atmosphere (see pages 26 and 62), but they also helped shape the planet's orbit, leaving Mars uniquely susceptible to the long-term cycles that affect both its climate and its geology (see page 66).

Internal structure

Like all planets, Mars is differentiated: gravity pulls it into a spherical shape, while internal heat makes the rocks mobile and allows them to separate according to their density. As a result, heavier material sinks to the core while lighter minerals rise to form a mantle and crust.

The size of a planet determines both the amount of heat generated during its formation and its ability to retain that heat—all other things being equal, the larger the planet, the longer it will remain hot. The interiors of small objects in the inner solar system have long since cooled and solidified, while those of larger planets (most obviously Earth, but probably Venus as well) remain hot with molten cores. This leaves the structure of Mars somewhere in the middle.

Investigating the deep interior of any planet is a challenge—on Earth, geologists can at least use the propagation of seismic waves from earthquakes as a means of mapping the inner layers, but on Mars this is not yet possible, so we can only rely on theoretical models and external measurements. The orbital periods of the Martian moons Phobos and Deimos can be used to calculate the planet's overall mass and therefore its density, which at around 0.14 pound per cubic inch (3.93 g per cubic cm) is just 70 percent of Earth's. More recently, accurate "geodetic" data from orbiting artificial satellites has been used to map the distribution of mass inside Mars and around its surface.

These measurements suggest that the internal structure of Mars is somewhat different from that of Earth. Its core is 2,230 miles (3,590 kilometers) across—almost half of the planet's overall diameter —but is lighter and less dense. While Earth's core is believed to be a mix of iron and nickel, the Martian core is probably composed of iron with roughly 16 percent of the light element sulfur. Above this lies a mantle of silicate minerals similar to those making up much of Earth's interior, but perhaps richer in iron. This is determined by the composition of Martian surface rocks and meteorites, which ultimately originated as volcanic magma formed in the mantle.

The precise nature of the Martian core, however, remains open to question. One big clue comes from the planet's magnetism, or rather the lack of it— Mars today does not have a strong global magnetic field like Earth's but the iron in its surface rocks preserves weak "fossil" magnetism—patterns of alignment in magnetic particles within the rocks that show that a global field was present as they solidified from their initial molten state. Until the early 2000s, there was general agreement that Mars had Earthlike magnetism from a molten core in its ancient past, but that the magnetic field had faded as the core cooled and solidified.

In 2003, however, geodetic measurements from the Mars Global Surveyor spacecraft (see page 178) threw doubt on this neat model. This satellite's instruments were sensitive enough to measure small-scale flexing in the Martian surface due to tidal forces as the planet went through its daily rotation. Lacking a substantial Moon or liquid oceans, Martian tides are much less impressive than those on Earth, amounting to little over 0.4 inch (1 cm), but this was still more than would be expected if the planet was entirely solid. The tidal evidence suggested that a liquid core still survives deep inside the planet, greatly adding to its overall flexibility.

Subsequent laboratory experiments have studied the behavior of the proposed iron/sulfur mix under the extreme pressures and temperatures of the Martian interior, and supported the general idea of a liquid core. The iron/sulfur blend has a significantly lower melting point than Earth's mixture of iron and nickel, and this, combined with the heat from decaying radioactive elements in Martian rocks, may have kept the Martian interior warmer than expected over the 4.5 billion years since it formed. The lack of a strong magnetic field remains a puzzle, but according to some models it could be explained if the Martian core, unlike Earth's, is entirely liquid— our planet has a molten outer core and a solid inner core, and this arrangement may play a crucial role in the "dynamo" process that generates our planet's magnetism.

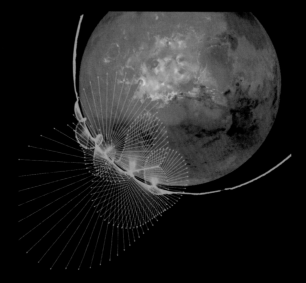

← Measurements carried out using Mars Global Surveyor's magnetometer experiment confirmed that the Martian magnetic field is weak and jumbled, with individual areas of the planet showing their own fields, but no overall "dipole" pattern of magnetic poles as seen on Earth. This is strong evidence that there was once a magnetic dynamo operating inside Mars, but it has long since ground to a halt.

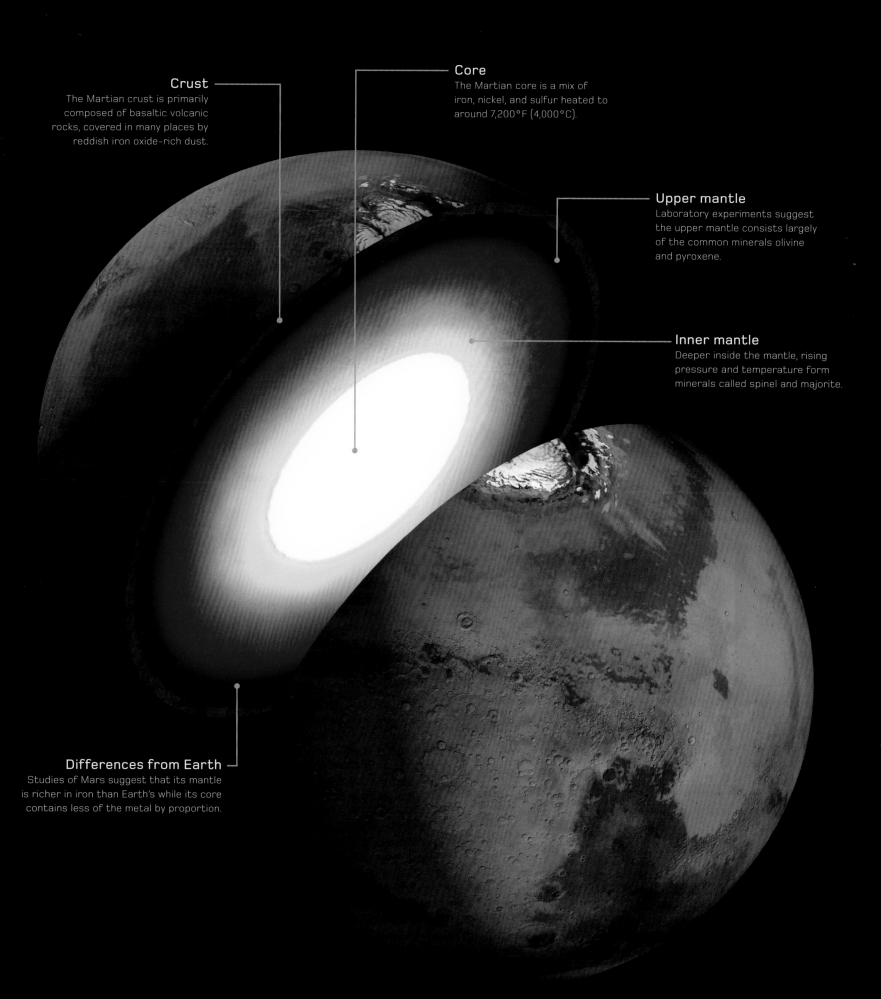

Crust
The Martian crust is primarily composed of basaltic volcanic rocks, covered in many places by reddish iron oxide-rich dust.

Core
The Martian core is a mix of iron, nickel, and sulfur heated to around 7,200°F (4,000°C).

Upper mantle
Laboratory experiments suggest the upper mantle consists largely of the common minerals olivine and pyroxene.

Inner mantle
Deeper inside the mantle, rising pressure and temperature form minerals called spinel and majorite.

Differences from Earth
Studies of Mars suggest that its mantle is richer in iron than Earth's while its core contains less of the metal by proportion.

Evolving world

Scientists divide the complex history of Mars into four broad periods, similar to Earth's own geological ages. As with other planets of the solar system, these ages are named in reference to major features that formed at the time.

The Pre-Noachian period lasted from the formation of Mars itself around 4.5 billion years ago, through to the formation of the huge Hellas impact basin, around 4 billion years ago. Most of the features that formed at this time have subsequently been destroyed or buried by later erosion, but one major event left a lasting impression that can still be seen today. This is the "crustal dichotomy"—a marked difference between the elevated, heavily cratered southern highlands and the much smoother northern lowland plains. The cratering record clearly shows that the southern hemisphere was exposed to the major impacts of the so-called Late Heavy Bombardment period around 4 billion years ago (see page 23), while for some reason—most likely because it was still essentially molten—the northern half of the planet remained unaffected. Several explanations have been proposed for this difference, but the most likely are either large-scale recycling of the Martian crust in the northern

hemisphere or one or several major impacts resurfacing the crust, either directly or indirectly by inducing later volcanic eruptions. Two other major basins—the Argyre basin in the southern highlands and the Isidis basin on the edge of the northern plains—also formed in the Pre-Noachian.

The ensuing Noachian period is named after the Noachis Terra region of the southern highlands. It lasted from perhaps 4.1 billion to 3.7 billion years ago (with no rock samples available for radiometric dating, it's impossible to be more precise with Martian dates). Geologically speaking, however, it forms a distinct "system" characterized by heavy levels of impact cratering and the likely presence of large amounts of water on the Martian surface, which formed standing bodies of water and carved river channels.

↓ Mars originated as a coalescing ball of molten rock around 4.5 billion years ago (1). During the pre-Noachian, the planet was heavily bombarded from space, while volcanism created a thick atmosphere and allowed water to accumulate on the surface (2). The Noachian period was marked by major impacts during the Late Heavy Bombardment (3).

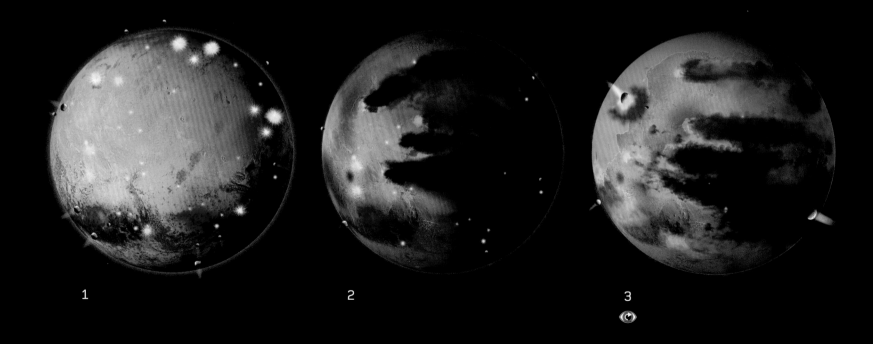

1   2   3

The next phase of Martian evolution, known as the Hesperian, takes its name from Hesperia Planum, a lava plain in the southern highlands. Such lava plains are a characteristic feature of this period, suggesting widespread volcanic activity, and are usually dated between 3.0 billion and 3.7 billion years ago. This is the period in which Olympus Mons probably began to form on the preexisting Tharsis rise. The beginning of the Hesperian also seems to coincide with a long-term change to Martian climate. Scientists are not yet sure of how Mars lost the bulk of its atmosphere, but possible processes involved include gradual erosion by the solar wind (the stream of particles blowing out from the Sun itself), absorption into the surface rocks as a result of catastrophic runaway climate change, or ejection into space by one or more large meteorite impacts. Whatever the cause, as the atmosphere cooled and thinned, standing bodies of water became ephemeral and much of the water on Mars

either froze or retreated below ground, from where it could still emerge in occasional catastrophic floods that left huge scars in the landscape (see page 54).

The latest phase of Martian history, the Amazonian, began around 3 billion years ago and extends to the present time. It is named after Amazonis Planitia, a broad lowland lava plain separating the Tharsis and Elysium volcanic regions, thought to have formed from as recently as 100 million years ago. The Amazonian is characterized by relatively low levels of impact cratering, but it covers such a huge span of time that it is difficult to talk about the dominant surface processes at work. Nevertheless, it's clear that throughout the Amazonian, glaciers played a great role in shaping the Martian surface, while flows of liquid water have become less important as the planet has dried and cooled, and major volcanic eruptions have also become less common.

↓ Widespread volcanism continued during the Hesperian period (4), but water retreated beneath the surface as the atmosphere thinned. Occasional catastrophic floods, however, continued to sculpt the landscape. During the Amazonian, ongoing volcanism and occasional flooding have continued to shape the surface of Mars (5), though the planet has settled into the generally cold and dry state we see today (6).

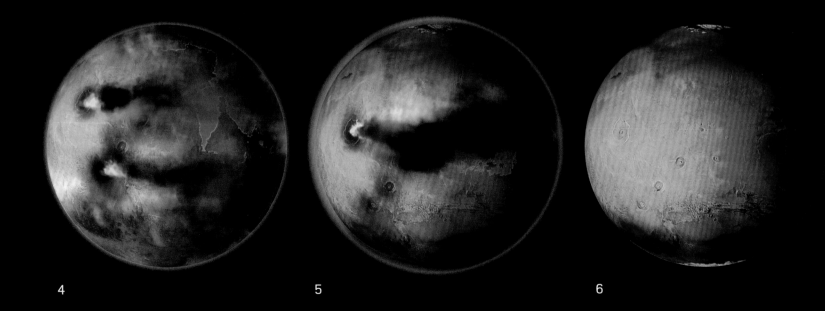

4   5   6

Structure of the crust

The Martian surface displays many signs of a complex history, shaped by changes to both geology and climate. One of the planet's most striking features is the marked difference or crustal dichotomy between the northern and southern hemispheres.

Broadly speaking, the planet's southern half consists of heavily cratered highlands while its northern hemisphere is composed of smooth lowland plains. This difference is more than simply skin-deep—locations in the southern hemisphere are typically 0.6–1.9 miles (1–3 kilometers) higher than those in the north, and the southern crust is far thicker, an average of 36 miles (58 kilometers) deep compared to 20 miles (32 kilometers) beneath the northern plains.

The dichotomy is powerful evidence for a cataclysmic event early in Martian history— either one or several major collisions late in the formation of Mars or some internal trauma that triggered a rapid resurfacing or recycling of the original northern-hemisphere crust. The deeper crust beneath the southern hemisphere may have developed as an aid to buoyancy, supporting the additional weight of the raised uplands just as the submerged part of an iceberg supports the part sticking out of the sea.

Both sides of the dichotomy are ultimately volcanic in origin—formed from molten magma building up in layers on the surface and dominated by igneous (volcanic) rocks. Sedimentary rocks are rare, and

apparently concentrated in a few specific regions such as Arabia Terra (see page 114).

Instruments such as the Gamma-Ray Spectrometer carried aboard NASA's 2001 Mars Odyssey space probe have now mapped the mineral composition of the planet's surface in some detail and reveal that the dichotomy makes its presence felt here, too. The southern highlands are dominated by minerals that are rich in magnesium and iron, including olivine, pyroxene, and plagioclase feldspar. This composition makes them very similar to volcanic basalts found on Earth and the Moon. Pyroxenes around the major volcanic shields tend to be richer in calcium, which would allow them to remain molten at lower temperatures. The volcanoes are thought to be considerably younger than the southern highlands, and this is evidence that Martian magmas have grown somewhat cooler over the planet's history.

The composition of the northern plains, meanwhile, is less certain, thanks in large part to the dust that blankets them. Exposed dark areas such as Syrtis Major appear to contain similar basalts to the highlands, but elsewhere the surface seems to be richer in silicates, which geologists believe suggests they could have formed from the eruption

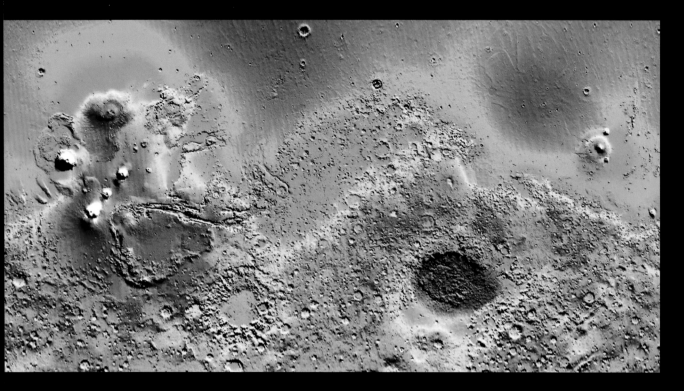

← This topographic map of the Martian surface is based on measurements from the MOLA laser altimeter experiment carried on NASA's Mars Global Surveyor spacecraft. Colors ranging from blue through green to yellow, orange, red, and white indicate increasing altitude and clearly reveal the difference between the two Martian hemispheres. Height differences around the average surface datum are far more striking than those on Earth.

Lava plains

In many places, the original Martian crust has been covered by relatively young plains of solidified lava.

← This artist's impression of a section through the Martian outer crust reveals a layered structure with hypothetical water-bearing aquifers and deeper layers of frozen ice created by the retreat of ancient surface waters. Although the evidence for deformation by tectonic forces is still uncertain, rocks have been transformed in places by shockwaves from meteorite impacts, leaving the upper layers in a jumbled mess known as "regolith."

↓ This view of a crater on the floor of Eos Chasma (part of the deep Valles Marineris canyon complex) reveals a mix of Martian bedrock materials. The HiRISE camera on Mars Reconnaissance Orbiter shows eroded basalt, long exposed to the surface, in red, and fresher rock revealed by the impact in blue and white. The brown dust in the middle of the crater (at center left) was probably blown into the crater by wind long after its formation.

Ice pockets

Ice permeates much of the crust, perhaps melting in some places to form liquid water aquifers.

Deeper layers

The oldest and deepest parts of the crust are composed of jumbled impact debris, or "regolith."

of magmas that were richer in low-melting-point, volatile chemicals. Another theory, however, is that the silicate-rich layer is a thin coating of glass and other minerals formed in the presence of water.

One of the most intriguing questions about the Martian crust, however, is its large-scale structure —is it an all-encompassing solid shell around the planet, or does it shift, crack, and rearrange itself in the same way as Earth's? The evidence is contradictory. On the one hand, Mars should have less internal energy to drive a system of Earthlike "plate tectonics," it has few signs of obvious plate boundaries and some geologists think the Martian shield volcanoes were only able to grow so big because heat from the interior had nowhere else to go. But on the other hand, the chainlike distribution of the Tharsis volcanoes suggests a long, slow drift over a "hot spot" in the mantle (see pages 30 and 90) and geologists have recently found evidence of significant "Marsquakes" that may be the result of recent tectonic activity.

A 2012 study that set out to answer this question once and for all came to a surprising conclusion. An Yin, a planetary geologist from the University of California at Los Angeles, identified fault patterns in the Martian crust that could only be a result of tectonic movements—most intriguingly, a 93-mile (150-kilometer) lateral offset between opposite sides of the Valles Marineris canyon system. Tectonics on Mars may not be very well developed, with just a few major plates shifting every million years or so rather than the several dozen in constant motion that our planet boasts, but the fact that they are present at all has important implications for the planet's past and future.

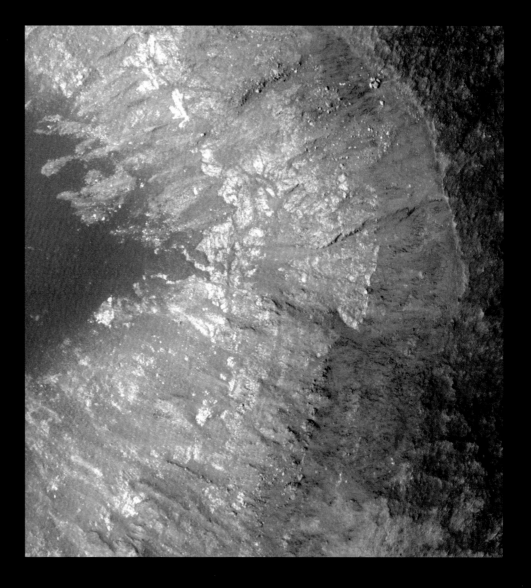

Martian volcanoes

Volcanoes are the most prominent and famous of all Martian features, thanks largely to the presence of the towering Olympus Mons, a giant volcanic shield that is the tallest mountain in the solar system. However, Martian volcanism comes in a huge variety of forms.

The largest shield volcanoes cluster on the huge bulge in the Martian surface known as the Tharsis rise. Ranging up to 6 miles (10 kilometers) above the average Martian surface level or "datum" (equivalent to sea level on Earth), the rise supports both the enormous Olympus Mons and a chain of three slightly smaller volcanoes—Ascraeus, Pavonis, and Arsia Montes—and is a puzzle in itself (see page 90). This volcanic chain bears an irresistible resemblance to similar chains found on Earth, such as the Hawaiian Islands, and probably has a similar cause—a hot spot of upwelling material in the mantle triggering activity that gradually shifts location relative to the surface. On our planet, this movement is understood to be the result of mobile tectonic plates in the crust moving slowly over a long-lived, static "mantle plume." However, the existence of a similar phenomenon on Mars, whose crust is normally assumed to be fixed, has been problematic (though see page 29).

Shield volcanoes form when pressure from a reservoir of hot magma deep below the crust is eventually released as lava escapes through a series of fissures. Over long periods, solidified rock builds up in layers to create a broad, shallow dome, with lava forcing its way through cracks in the flanks. Eventually, when the magma reservoir beneath is exhausted, the volcano's peak may collapse to form a broad crater or caldera, while steep areas on the flanks may give way to form clifflike escarpments.

Other large-scale volcanoes on Mars are known as tholi (from a Greek word for a building with a domed roof) and paterae (from a Latin word for a broad, shallow bowl). Tholi are steep-sided and have relatively large calderas for their size, with a superficial resemblance to classic cone-shaped

stratovolcanoes on Earth, such as Japan's Mount Fuji and Italy's Mount Vesuvius. However, appearances are misleading and geologists now believe they are the exposed peaks of much broader and flatter shield volcanoes that have been largely buried beneath later lava flows.

Paterae are completely different—shallow, ragged-edged "scabs" on the Martian surface that probably mark the sites of intermittent violent eruptions rather than the long-lived and relatively sedate volcanism of the shields. They can be huge in extent (even larger than Olympus Mons), but are much flatter. Some extremely ancient paterae may have qualified as fully fledged "supervolcanoes" in the distant Martian past (see over).

Needless to say, one of the most intriguing questions about Mars is whether it remains volcanically active today. This has yet to be answered for certain—scientists have not yet seen an eruption in progress, but they have found suggestive evidence. Some areas on the flanks of Olympus Mons appear so fresh (with little sign of the impact cratering geologists use to date planetary surfaces) that they were probably created just 20 million years ago. Active volcanoes are also one potential source for the controversial reports of methane in the Martian atmosphere (see page 74). In 2005, meanwhile, the European Space Agency's Mars Express mission discovered fields of small volcanic cones in the Martian arctic, each just a few hundred feet high and in such pristine condition that they were probably active up to 2 million years ago, if not more recently. If Mars is able to sustain such activity today, then it is clearly not the geologically dead world that scientists have long assumed.

↓ A shield volcano begins to form when a huge intrusion of magma, heated by a hot spot in the underlying mantle, pushes the crust upward to create a series of volcanic vents. (1) Lava erupts from these vents and starts to build up in layers (2) that thicken over time, as the lava magma finds new ways to the surface (3). Eventually, the magma reservoir is exhausted or solidifies as the hot spot moves on, allowing the summit to sink and form a caldera, while the flanks slump outward (4).

1 **2** **3** **4**

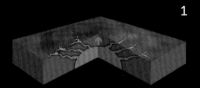
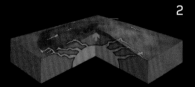
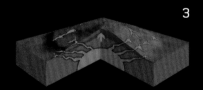
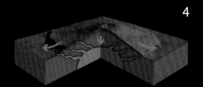

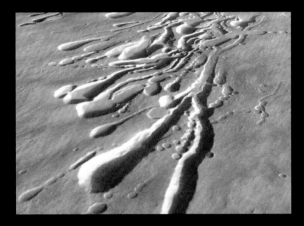

↑ A three-dimensional view from the Mars Express mission reveals a network of tadpole-shaped channels on the flank of the huge Pavonis Mons shield volcano. They are thought to be collapsed lava tubes—remains of ancient tunnels through which molten lava flowed just beneath the volcano's solid surface. Long after the eruptions came to an end, the tunnels of the roofs caved in, forming the U-shaped channels we see today.

→ Apollinaris Mons is an ancient shield volcano on the edge of the Martian southern highlands. Its summit rises to 3.1 miles (5 kilometers) above its surroundings, and is topped by an 50-mile (80-kilometer) caldera. In this false-color image from ESA's Mars Express space probe, the blue areas highlight clouds that have formed in the thin air above the enormous peak.

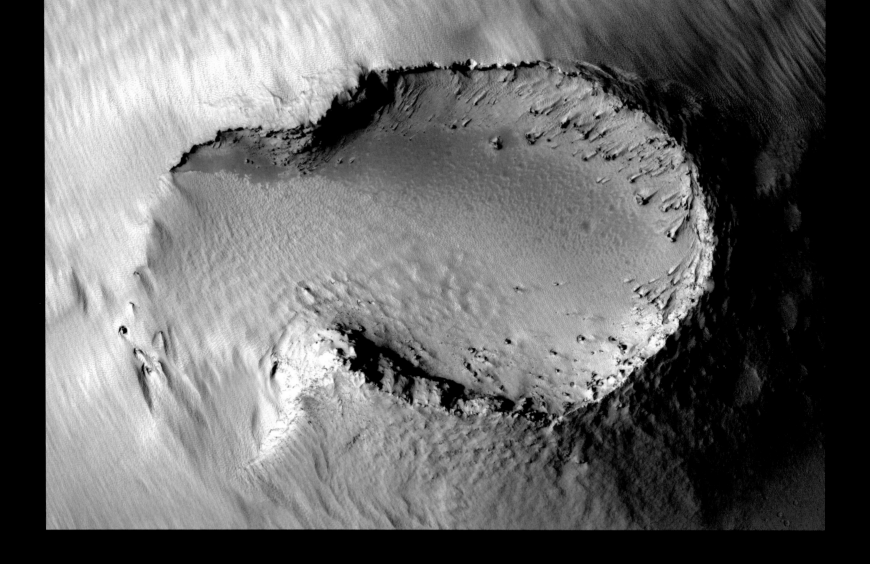

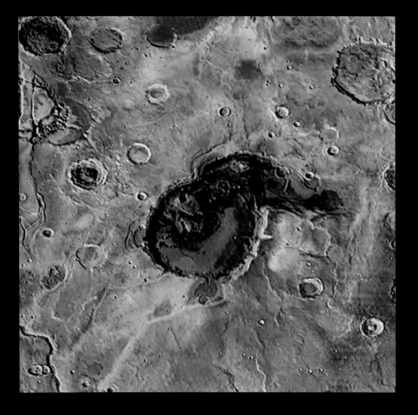

← This color-coded elevation map from Mars Reconnaissance Orbiter reveals an enormous sunken pit in the Arabia Terra region (see page 114), known as Eden Patera. It is thought to mark the site of an ancient supervolcano, similar to those found on Earth at Yellowstone National Park in the northwestern USA and Lake Toba in Sumatra. In contrast to relatively sedate shield volcanoes, supervolcano eruptions are sudden and immensely violent, with potentially catastrophic effects for a planet's climate.

↑ Many Martian volcanoes are so big that their individual volcanic vents take on features that we would recognize as separate volcanoes on Earth. This image from NASA's Mars Reconnaissance Orbiter, for example, highlights a conical structure on the flank of the enormous Pavonis Mons volcanic shield. This is probably a cinder cone—a pile of volcanic rock roughly 2,300 x 3,600 feet (700 x 1,100 meters) across, formed by a short-lived fountain of lava.

→ A colorful elevation map reveals features of Tharsis Tholus, an ancient shield volcano on the Tharsis rise that has been modified by extensive faulting and shows two distinct calderas. While the exposed shield is 96 x 78 miles (155 x 125 kilometers) in diameter, more extensive outer flanks are thought to lie buried beneath a younger surrounding lava plain.

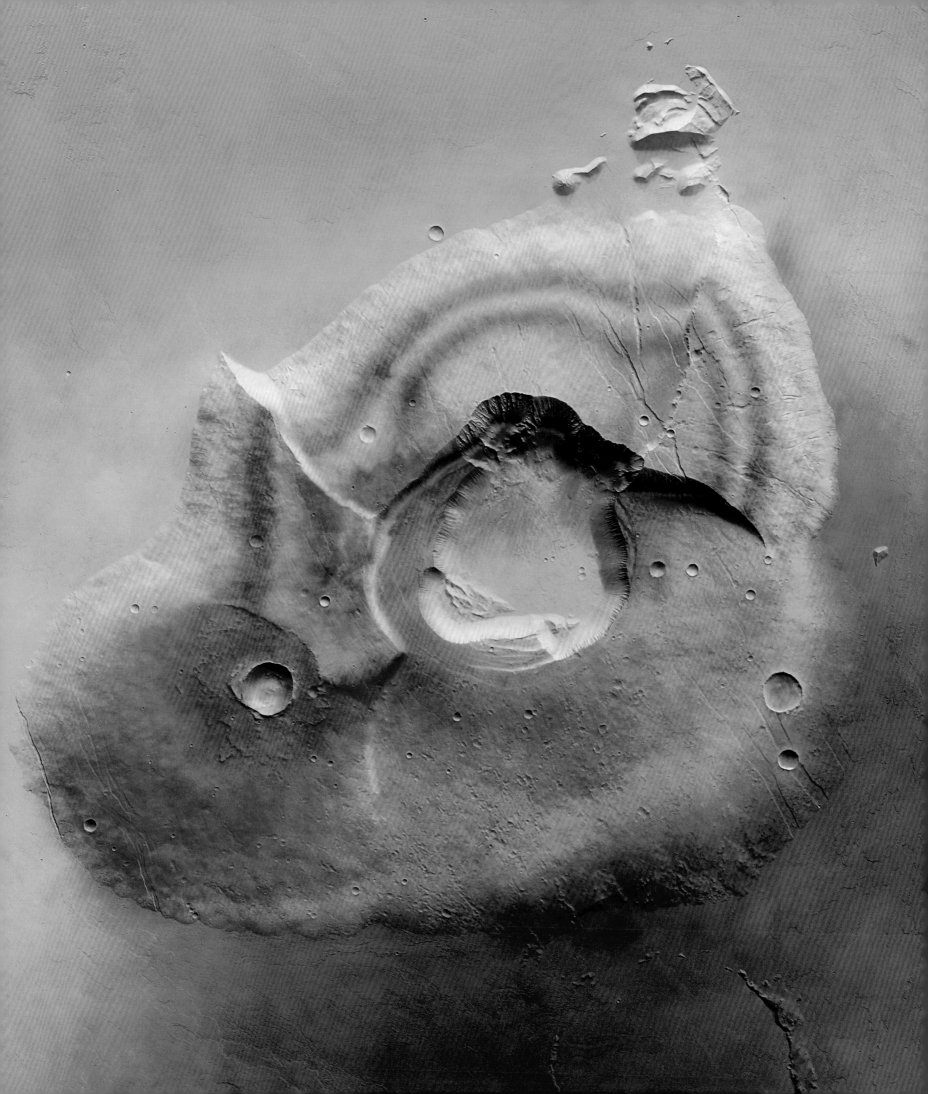

Sands of Mars

Mars gets its famous color from the fine, ruddy-colored sands that cover much of its surface. These are the major component of the Martian soil—a term that encompasses all small particles on the surface and does not imply any organic content.

Measurements taken by landers and rovers in several different areas of the planet have shown that the sand is essentially the same across the globe—as perhaps might be expected given the enormous dust storms that frequently stir it up and redistribute it. In general, the sands are much finer than most of those found on Earth, with average particle diameters of just a few tens of micrometers (millionths of a meter)—closer to the size of dust. The size difference is due to the sustained period in which erosion has been a dominant force on the Martian surface, gradually wearing down individual particles, and the lack of present-day water or other fluids that encourage individual grains to stick together and grow in size.

The smallest component of the soil, with particle diameters of around 3 micrometers, is classed as Martian dust. Laboratory simulations of the atmospheric conditions and models of the effects of low Martian gravity suggest that grains as large as 20 micrometers can be carried into the air, where they may remain suspended for long periods —and certainly a pinkish pall of dust is a permanent feature of the Martian sky even between the major dust storms.

Closer to the surface, Martian winds also act on the larger sand grains in a variety of ways. Just as in Earth's own deserts, gusts of wind near the surface can trigger a runaway process called saltation, in which sand grains cascade across the ground on ballistic trajectories. The tiny impacts of individual grains shake others loose so that they too are carried on the wind and the process repeats rather akin to a small-scale avalanche. The major difference between Earth and Mars is that the low air pressure and gravity combine to make the individual trajectories of saltating particles much longer and flatter. Saltation is a major factor in the instigation of dust storms, and also helps build and shape the many different kinds of sand dune that cover flat areas of the Martian surface.

Chemically, both dust and sand get their reddish color from the presence of iron oxide, better known as rust. The most popular explanation for this chemical's abundance on Mars is that the planet's crust is enriched with greater amounts of iron because its interior did not differentiate so thoroughly as Earth's early in its history (see page 24). While much of Earth's iron grew hot enough to melt and sink to the core of our planet, a great deal of Martian iron remained at the surface, where it

subsequently reacted with surface water during the wet Noachian phase of Martian history. However, this is not the only possible explanation for Mars's appearance—alternative methods for rusting the planet without the need for abundant water have been put forward. Such explanations for the reddish color rely purely on the erosion and mixing of sand and magnetite from the planet's widespread basaltic rocks.

In fact, the Martian soil must ultimately have derived from just such sources, whether they explain its color or not, and the Curiosity rover's recent analysis of Martian soil has identified many of the same mineral components—feldspar, pyroxenes, and olivine—previously found by the 2001 Mars Odyssey orbiter in highland rock outcrops (see page 180). However, it would be a mistake to assume that the Martian soil is identical in all places. New discoveries, such as magnesium and potassium nutrients found in the Martian permafrost by NASA's Phoenix lander (see page 200) and silica-rich "hot spring" material discovered by the Spirit rover, have revealed considerable complexity and local variation.

→ Curious dark streaks on the Martian surface, shown here in a valley system known as Acheron Fossae, are thought to be formed when fine-grained sand flows down relatively steep slopes and exposes the darker volcanic sands beneath. A similar process takes place when dust devils scour the landscape (see page 70).

↓ Two different scales of sand structure combine to form a strange and beautiful landscape in Proctor crater. Small ripples are thought to be formed from fine-grained sand and are relatively slow-moving, so they accumulate a coating of bright dust over time. Larger dunes, in contrast, are made from darker volcanic sand, and migrate across the surface more rapidly so they do not accumulate a brighter surface layer of dust.

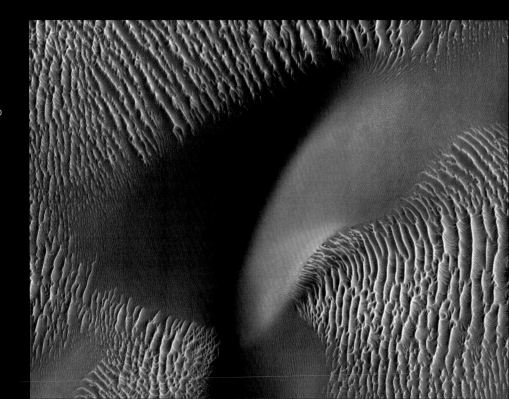

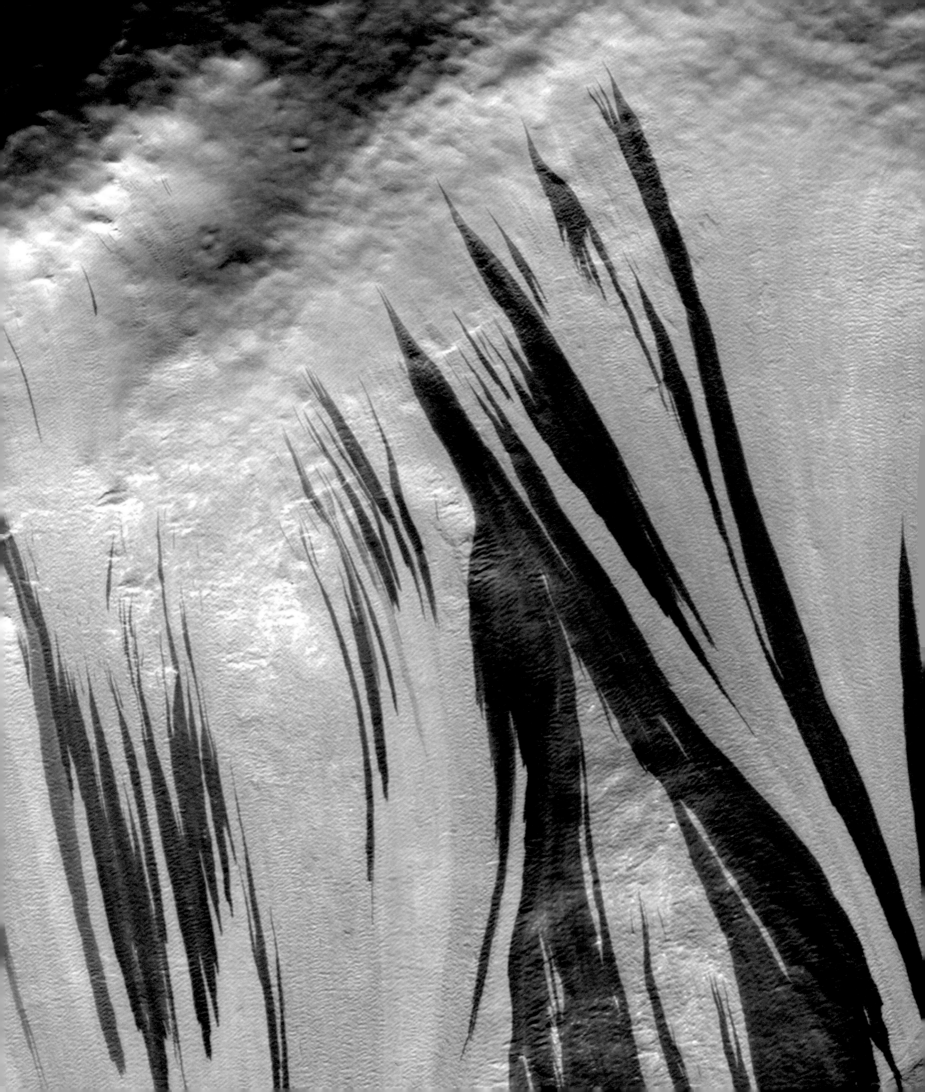

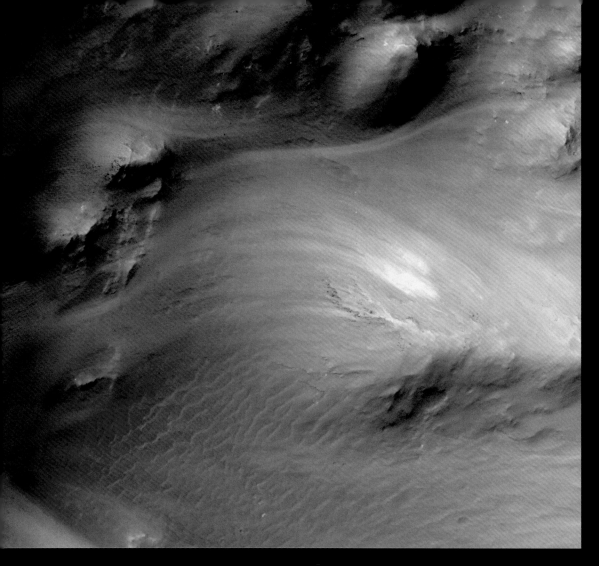

↳ This surreal image shows complex sand patterns that have formed on a crater floor in the Noachis Terra region. The entire region of the picture is around 0.6 miles (1 kilometer) wide, so the image captures features on scales down to just 40 inches (1 meter) or so across.

← This enhanced-color view reveals complex patterns around the centre of an unnamed highland impact crater. Bedrock uplifted to form a central peak has subsequently slumped back, and is now eroding to create beautiful streams of fine-grained sand.

↓ Another crater floor in the Noachis Terra region gives shelter to the distinctive and beautiful structures known as barchan dunes. Despite their strange appearance, these dunes arise naturally in the presence of just the right amount of sand and sustained winds from a single direction. This false-color image from the Mars Reconnaissance Orbiter includes an infrared (heat radiation) channel that would normally be invisible to our eyes. Another complex field of barchan dunes, within a large impact basin known as Herschel, is shown overleaf.

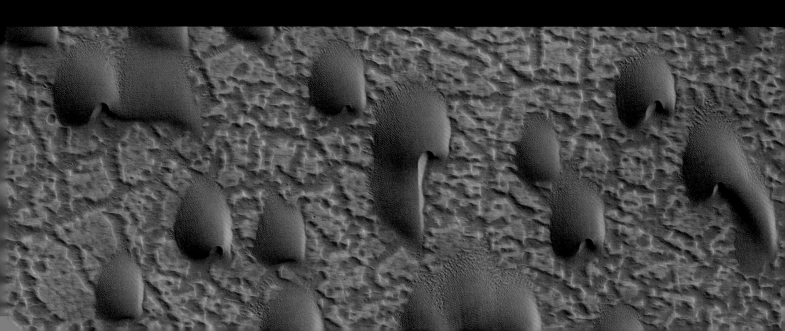

↑ NASA's Spirit Rover (see page 188) captured this beautiful mosaic, showing a field of rippled sand nicknamed "El Dorado," over several days around New Year 2006. The resulting image covers a 160° field of view, roughly approximating what an astronaut would see from Spirit's location. Satellite images show the El Dorado dune field as a notably dark patch on the side of Husband Hill within Gusev crater, and Spirit's analysis confirmed the presence of dark, rounded grains of sand originating from the erosion of volcanic rock.

Polar caps

The north and south poles of Mars are covered by bright ice caps, eerily reminiscent of Earth's own polar caps transplanted to an arid alien environment. These prominent areas were the first surface features to be spotted by early telescopic observers, and have fascinated astronomers ever since.

Part of the reason for their continued interest is the fact that the polar caps are constantly changing—waxing and waning in a cycle driven by the Martian seasons. What's more, the structures of the two caps, and their patterns of seasonal behavior, are very different from each other. Each cap is thought to be composed of a layer of permanent frozen water ice topped by an ephemeral sheet of frozen carbon dioxide that accumulates in the fall and largely disappears in spring.

The north polar cap is the less variable of the two—with a diameter of around 750 miles (1,200 kilometers) it covers the Martian arctic down to a latitude of about 80° more or less permanently and is surrounded by the vast plain of the Vastitas Borealis (see page 110). The fact that it was so large and persistent, despite the northern hemisphere's generally warmer seasonal conditions (see page 64), suggested early on that it contained substantial amounts of frozen water ice (which is more or less stable within the range of temperatures experienced at the north pole).

In contrast, the southern polar cap permanently covers a far smaller region (around 250 miles or 400 kilometers across), but grows much larger in winter, extending down to a diameter of around 1,250 miles (1,800 kilometers). Such a dramatic change is only possible through the condensation and sublimation of huge amounts of dry ice (frozen carbon dioxide) from and back into the atmosphere. (Sublimation is a direct transition from solid to gaseous form without an intermediate liquid stage.)

Measurements from orbiting space probes suggest that the polar caps accumulate up to 80 inches (2 meters) of frozen carbon dioxide each fall, which may even fall as snow. Cameras on the Phoenix lander (see page 200) saw snow in the air during the fall, although this sublimated before reaching the ground. In the winter, snow would be more likely to make it all the way to the Martian surface.

The processes of fall growth and spring contraction both leave their own traces in the landscape. As layers of ice are laid down, they trap dust within them, which remains behind when the ice sublimates in spring. What's more, during periods of global cooling, a little more ice is left behind each spring, resulting in layered deposits, several miles deep, surrounding and underlying each pole. At the moment, however, there is some evidence, in the form of so-called "Swiss cheese" terrain, that Mars

is actually warming and the poles are losing a little more of their permanent ice each year (see page 66).

The patterns of ice and dust at both poles show complex fingerprint-like patterns—spirals and whorls that are in fact deep, ice-free canyons etched out of the layered deposits. These are thought to be a result of prevailing winds and changing amounts of sunlight affecting the locations where ice tends to accumulate or sublimate more easily.

Each spring sees the ice field develop spectacular features known as starburst channels. These dark spots, surrounded by countless radiating spokes, are created where dark dust in the ice layer absorbs solar radiation and warms its surroundings. The result is a layer of gas trapped just beneath the surface, which eventually finds its way out in geyser-like jets that carry dust with them and scatter it over the nearby surface.

→ "Starbursts" (sometimes called "spiders," are features that form in the ice around the Martian south pole during southern spring. They are thought to be the result of subsurface pockets of carbon dioxide ice sublimating and breaking through the surface to create geyser-like jets.
↓ A Hubble Space Telescope view of Mars captures the region around the south pole at the height of summer. The polar region of the northern hemisphere, tipped away from Earth and Sun and only illuminated by weak sunlight, still shows signs of an extensive winter frost cap.

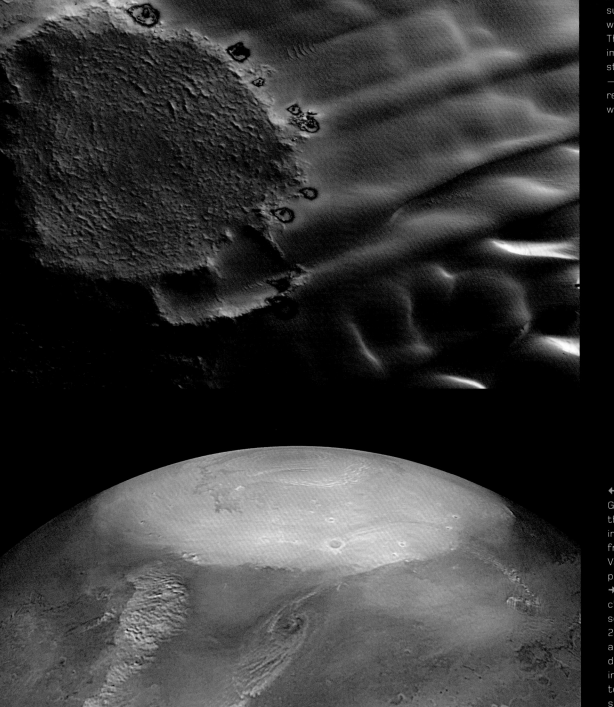

← Each year, dunes around the Martian north pole are covered by seasonal carbon dioxide frosts that sublimate back into the atmosphere with the onset of northern spring. This Mars Reconnaissance Orbiter image shows dunes that are just starting to lose their frost covering —their warmest sunlit slopes have released dark cascades of dust where the frost has disappeared.

← A spectacular image from Mars Global Surveyor shows much of the northern hemisphere in detail, including thick carbon dioxide frosts across the vast plains of the Vastitas Borealis, and the raised plateau of the polar ice cap itself.
→ This curious ring-shaped feature, close to the edge of the residual south polar ice cap is roughly 2.5 miles (4 kilometers) across. It appears to be a roughly circular depression surrounded and filled in by so-called "Swiss cheese" terrain—pits where material has sublimated from the layers of ice and dust, which have grown steadily larger and begun to connect with each other. The depression itself lacks a rim or central peak that might suggest it is an impact crater, so it is probably a result of subsidence from beneath.

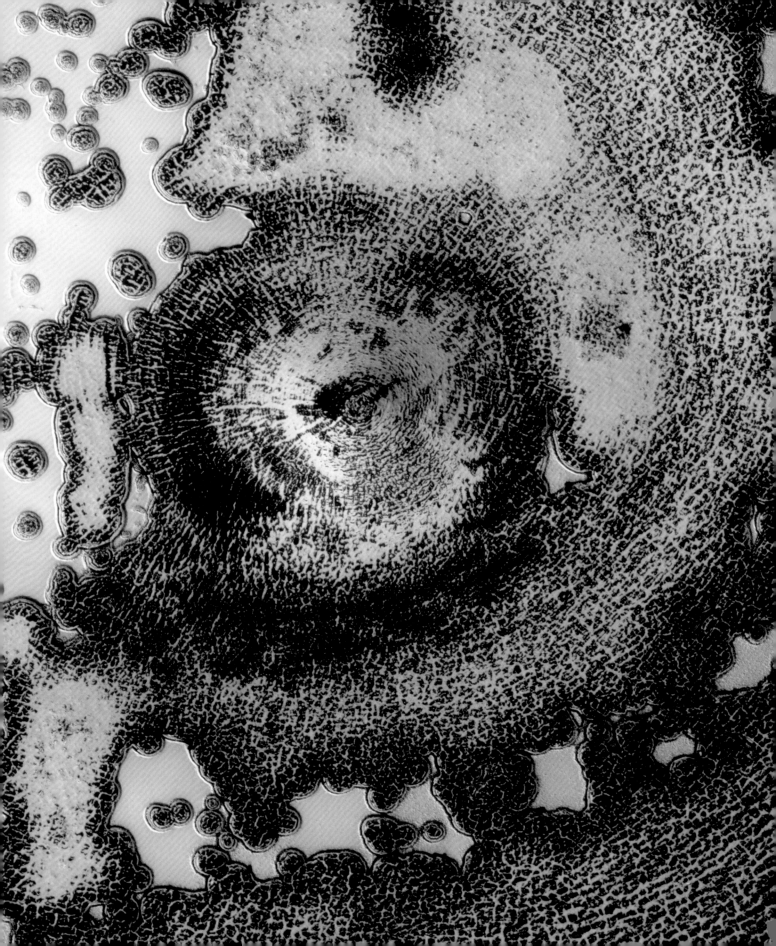

Subsurface ice

In 2002, NASA's Mars Odyssey spacecraft revealed the existence of vast amounts of hydrogen locked into the Martian soil at each hemisphere's middle and high latitudes (see page 180). This was powerful evidence that much of the Martian soil contains large amounts of water ice.

The existence of copious ice mixed with the Martian soile was subsequently confirmed by Phoenix, the first lander to target the hydrogen-rich area close to the north polar ice cap (see page 200). The discovery that Martian water ice extends well beyond the polar caps helps to explain a wide variety of landscape features imaged by orbiters such as Mars Express and Mars Reconnaissance Orbiter. In fact, it's becoming increasingly clear that much of the surface is more akin to dust-covered ice than merely icy dust. The dust covering insulates the ice beneath and helps protect it from sublimation at relatively warm latitudes where it could not otherwise survive.

The hidden ice often gives itself away through signs of its very gradual flow—under the influence of gravity or other factors, it changes its distribution, forming parallel grooves along the direction of its motion. Some Martian features are analogous to glaciers found on Earth, with a long, convex "river" of ice flowing down a valley, a crater wall, or even a volcanic slope, and spreading out at its snout to form a lobe that pushes rocky debris before it. In 2008, Mars Reconnaissance Orbiter used its radar to confirm the presence of subsurface ice and the glacial nature of these distinctive structures.

So-called "concentric crater fill," meanwhile, is rather different. It forms vast rippled plains that fill certain craters to the brim, making them far shallower than they should be for their size.

As tens, perhaps hundreds of feet of ice settle and compress under their own weight, geometric patterns develop naturally in the surface. Gradual sublimation when ice gets close to the surface can give rise to intriguing geometric landforms, whose evocative names such as "brain terrain" give some idea of their appearance. There is also a recognized effect known as terrain softening, whereby landscape features at latitudes poleward of 30° north or south appear rounded and smoothed compared to those around the equator—another result of ice gradually sublimating from the soil.

Although glacial features are found all around the planet at mid-latitudes and above, the most obviously ice-sculpted landscapes run along the dichotomy boundary, between northern plains and southern highlands, where they are known as "fretted terrain." These areas are marked out by broad, flat-floored canyons, tablelike mesas, and "knobs," and cliffs—often emerging from flat lowland plains. They are generally thought to have developed as a result of large-scale erosion—itself due to the sublimation of the ice-rich component from their rocks, but some areas still contain obvious signs of ice persisting today—not least the Ismenius Lacus region (see page 122), and the "frozen sea" in northern Elysium Planitia (see page 134).

While the planet's high-latitude ice may have persisted more or less intact for a billions of years, the ice closer to the equator has presumably got

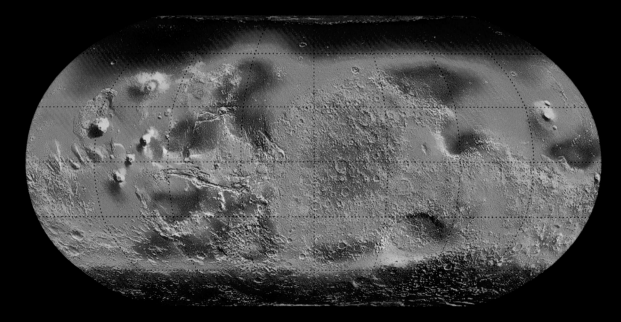

← This map, compiled using data from the 2001 Mars Odyssey probe, shows hydrogen-rich areas of the Martian soil in dark blue. Compiled using the spacecraft's Gamma-Ray Spectrometer instrument, the map reveals huge amounts of hydrogen, linked to the presence of water ice (frozen H_2O) in high-latitude permafrosts around the Martian poles.

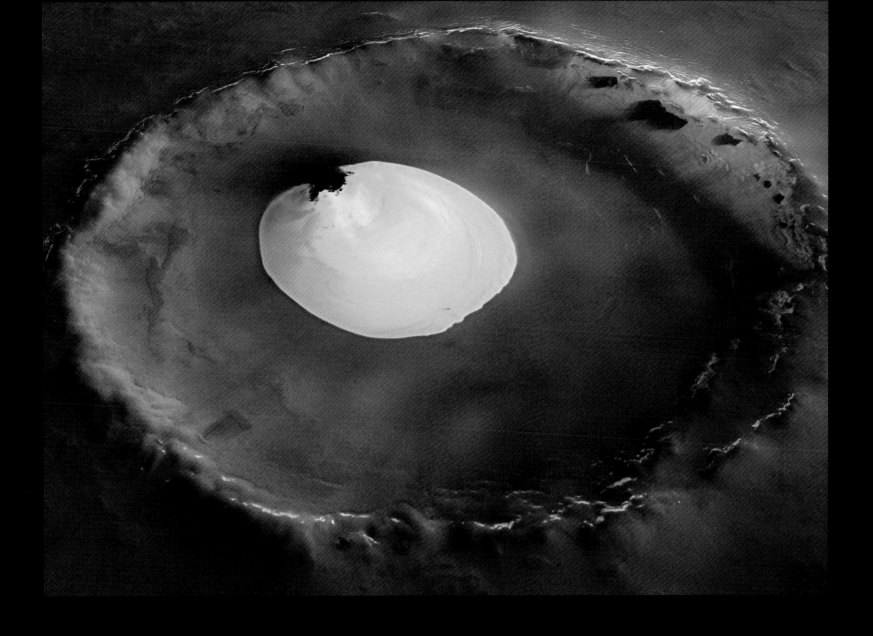

there more recently. Assuming the theories of Martian climate cycles are correct (see page 66), then some of the glaciers may have formed, like those on Earth, as a result of persistent snowfall accumulating year on year, during wetter periods of the Martian past. Another possibility, however, which certainly seems to account for some features, is that the ice was incorporated into the soil following catastrophic floods (see page 54). This would certainly help to explain the distribution of fretted terrain, since areas on the boundary between highlands and lowlands seem to be particularly vulnerable to such events.

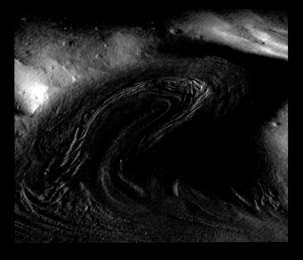

↑ This famous image from the High-Resolution Stereo Camera aboard the European Mars Express space probe shows a lakelike patch of exposed water ice sitting on the floor of a 22-mile (35-kilometer) crater near the Martian north pole.
← This 3-mile (5-kilometer) tonguelike lobe on the highland/lowland border has a startling resemblance to glaciers on Earth, and like other similar Martian features, is almost certainly a river of frozen ice disguised by a thin layer of dust and rocks.

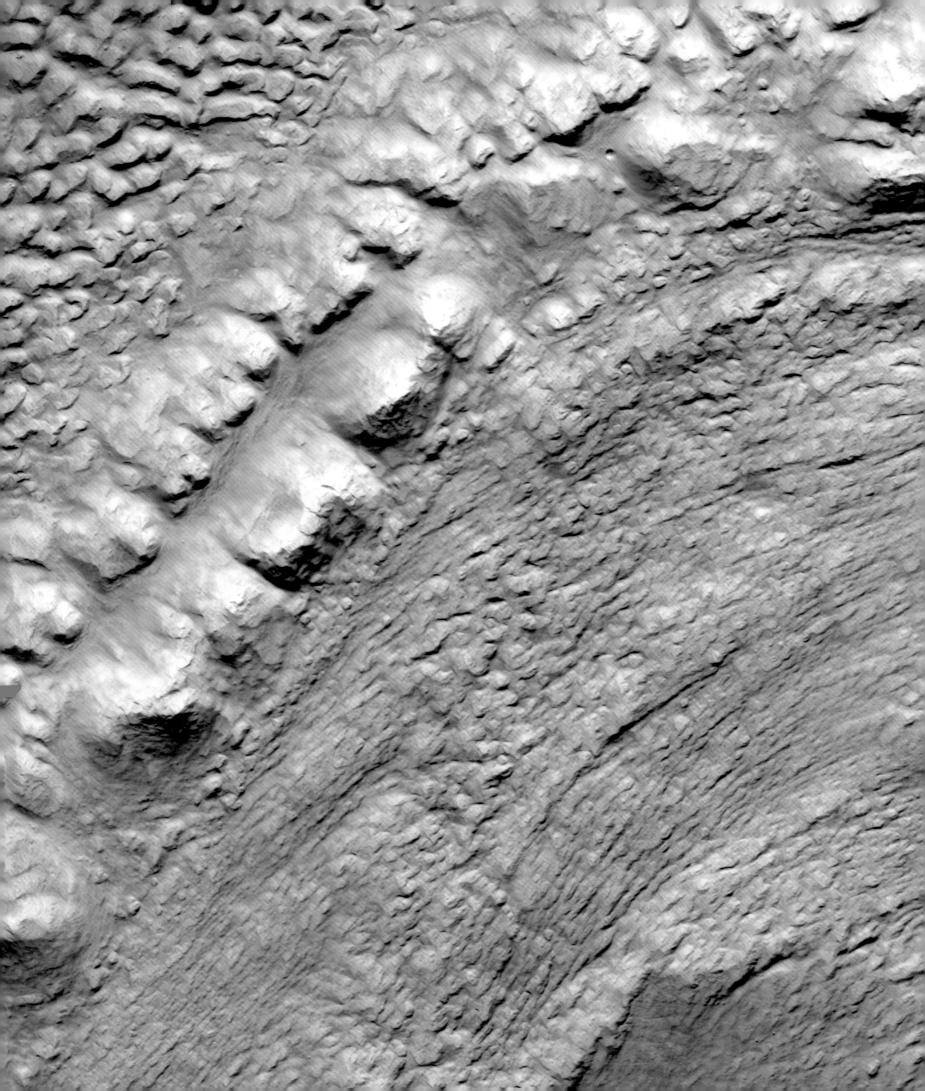

Protonilus Mensae is a rugged, ice-rich region in the mid-latitudes of the northern hemisphere. This false-colored image captures the telltale flow of a dust-covered glacier along a deep valley.

↓ This Mars Express image shows a fragmented, flattened landscape in Elysium Planitia (see page 134). Bearing an obvious resemblance to fragmented rafts of ice seen in Earth's polar seas, it is believed to be a frozen sea left behind by a catastrophic flood originating in the nearby Cerberus Fossae.

↓ This map of the Deuteronilus Mensae region was compiled using Mars Reconnaissance Orbiter's Shallow Subsurface Radar instrument (SHARAD). It confirms the existence of widespread and thick ice deposits hidden beneath thin layers of dust. At least some of this region's glacial features are thought to be very young, formed in the last 100,000 to 10,000 years.

The entire concept of impact cratering was controversial up until the 1960s, when early probes sent to crash-land on the Moon confirmed that our satellite is covered with craters at all scales from the huge down to the tiny, including many that are far too small to have been formed by volcanic activity. Nevertheless, impact cratering is now seen as a major force shaping the planets throughout the history of the solar system and a useful tool for dating planetary surfaces.

Thanks largely to lunar rock samples brought back by the Apollo missions, planetary scientists have a good understanding of the way that crater formation rates changed over the past 4.5 billion years. The precise amount of cratering at any particular time is dependent on the amount of small debris objects (a mix of rocky asteroids and icy comets—collectively termed meteoroids, or meteorites when they impact planets) orbiting in the vicinity of the planets. It seems that during the first few hundred million years after the formation of the planets, the number of meteorite impacts declined smoothly as the larger objects "soaked up" the smaller objects that crossed their path or ejected them into more distant orbits where impacts were less likely. Then around 4 billion years ago, the outward spiral of the giant planets Neptune and Uranus upset the cloud of small icy worlds known as the Kuiper Belt (see page 22), resulting in a sudden and dramatic spike in both the rate and size of impacts. Kuiper Belt Objects of all sizes were disrupted onto orbits that sent them plunging toward the inner solar system—an event known as the Late Heavy Bombardment.

The collisions resulting from this period include some of the largest craters in the solar system —the huge impact basins that were later filled in by volcanic eruptions to form the Moon's maria or lunar "seas," and Mars's equally impressive Hellas and Argyre basins. Earth, too, would have been struck by impacts on a similar scale, but the constant tectonic recycling of our planet's crust has wiped all trace of them from the geological record. It was at around this time that, according to one theory, Mars may have suffered an even more cataclysmic impact from a substantial body that all but obliterated its northern hemisphere, wiping out the record of previous cratering and ultimately allowing the formation of Mars's vast northern plains (see page 110).

By around 3.8 billion years ago, the Late Heavy Bombardment came to an end and cratering tailed off substantially—ever since, it has continued at a slowly declining background rate. Geologists can use this record of cratering to estimate the relative age of planetary surfaces: the principle is that the older a particular area of the surface, the more craters it will accumulate (assuming impacts on a particular planet happen more or less at random). Once calibrated with reference to other dating techniques, this method can be surprisingly precise.

Martian craters, then, can not only reveal the age of the different parts of the surface, but also offer other insights. The process of crater formation effectively gouges a chunk of material from the planet's surface, scattering debris or "ejecta" over the surrounding landscape, and also exposing deeper, older rock strata within the crater walls. The studies carried out by NASA's Opportunity rover at Victoria crater (see pages 128 and 192) have shown how useful such natural exposures can be. Even satellite studies of crater ejecta and the uplifted walls that typically surround craters of moderate size can provide more information about the nature of the Martian crust.

The shape of a crater, meanwhile, can also provide information about the nature of the meteor that formed it and the surface conditions where it struck—several Martian craters, known as rampart craters, show distinctive lobed ejecta patterns where material splashed out in mudflows as the crater formed. This allows geologists to identify locations where there was substantial ice mixed in with the soil that liquefied during the impact, or in the more ancient instances, where there may even have been bodies of standing water.

And there is one final result of these impacts for which scientists are grateful—very occasionally, an impact occurs with such force that ejecta rocks are flung free of Martian gravity to become meteoroids in their own right. Even more rarely, these crash on Earth as meteorites, where they provide us with unique samples of Mars rock for laboratory study (see page 74).

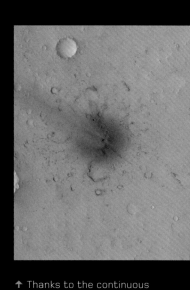

↑ Thanks to the continuous presence of spacecraft orbiting Mars since 1997, it's been possible to identify some extremely recent craters, such as this 79-foot (24-meter) example in Arabia Terra, formed by an impact at some point between April 2, 2001 and December 11, 2003. The crater is surrounded by streaks of dusty ejecta that erode rapidly and are not seen around older craters.

→ Galle is a large (140-mile/230-kilometer) crater in the southern highlands of Mars, on the eastern edge of Argyre Planitia (see page 154). It is famous as the "Happy face crater," with a smile and eyes formed by a smaller crater and internal mountain ranges.

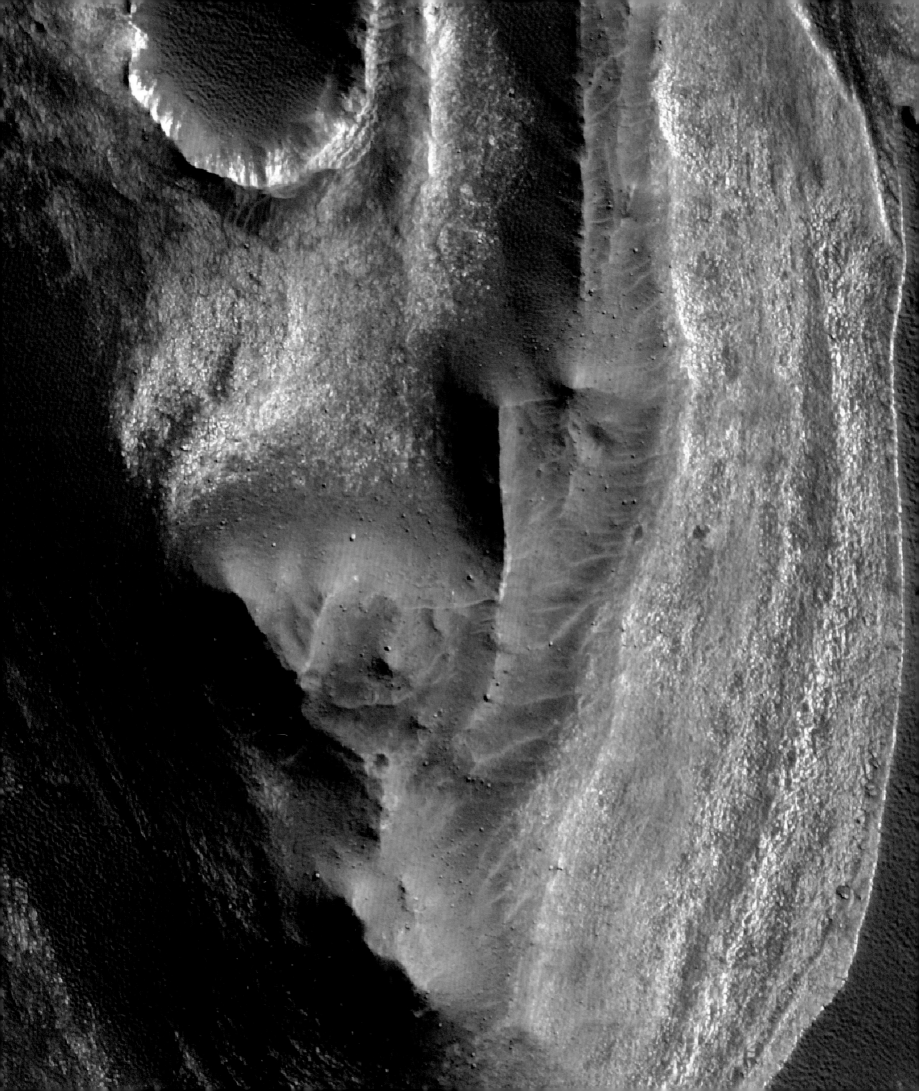

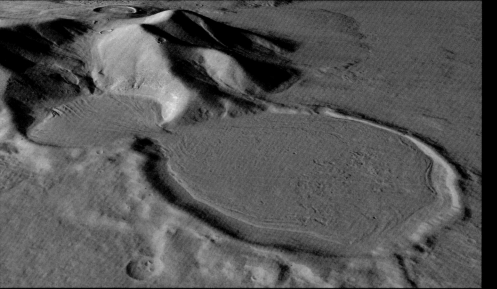

← This shallow hourglass-shaped crater on the eastern rim of the Hellas basin is a result of two craters, 5.5 and 10.5 miles (9 and 17 kilometers) wide, both being filled by a mix of glacial ice and rock that has flowed down from the massif seen in the background.

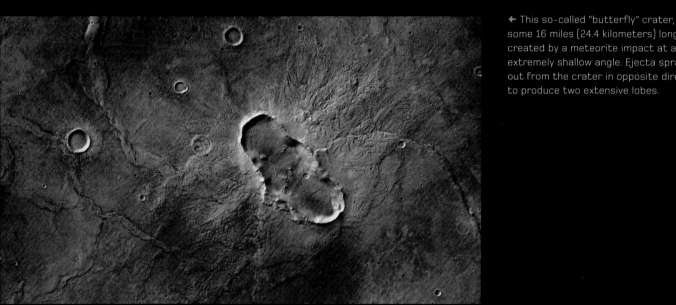

← This so-called "butterfly" crater, some 16 miles (24.4 kilometers) long, was created by a meteorite impact at an extremely shallow angle. Ejecta sprayed out from the crater in opposite directions to produce two extensive lobes.

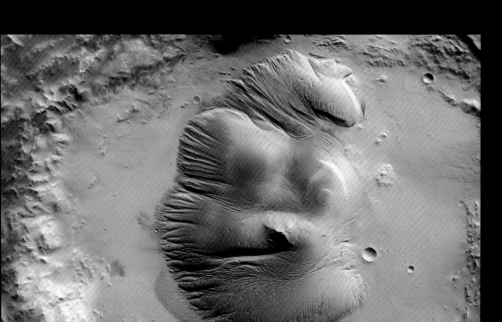

← Martian craters often contain substantial hills on their floors, as seen here in Nicholson crater on the edge of Amazonis Planitia. These hills seem to mark locations where fine-grained material has been deposited on the floor (either carried on the wind or during previous periods of flooding). The resulting sedimentary rock has subsequently been worn down by erosion.

←← This crater sits on top of an apparently insubstantial dune formation at the edge of Melas Chasma, part of the Valles Marineris canyon complex. Despite appearances, however, the shape of the crater reveals that the dunelike structure is solidified sandstone.

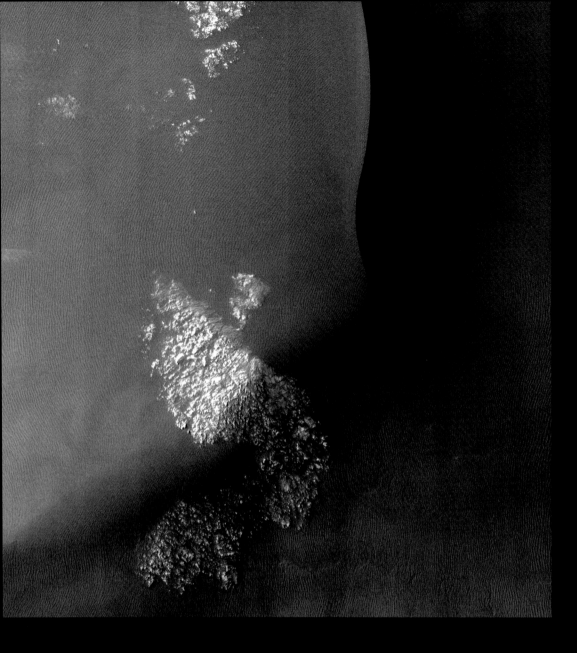

← The floor of this southern-hemisphere crater is filled by an extensive landscape of rippled, dark dunes surrounding a few bright outcrops of rock. At bottom left, lighter areas of the surface are painted with dark trails left behind by dust devil whirlwinds (see page 70).

→ Craters sometimes act as useful "windows" into deeper layers of the Martian crust. For example, a pit in the floor of this mid-sized crater in the highlands to the north of Hellas basin exposes ancient bedrock crisscrossed by dark, veinlike fracture patterns. The fractures, known as joints, are probably due to tectonic forces placing stress on the rock after its formation, while the dark material within them may be an accumulation of trapped dust. Near the bottom of the picture, the crater floor is entirely covered in this darker dust.

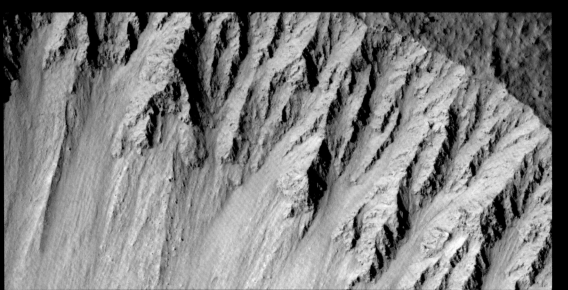

← The fine structure in the walls of this fairly young crater in the Tharsis region is unusual for Mars, and bears a stronger resemblance to that found in craters on Earth's Moon. This may suggest that the crust where the incoming meteorite impacted was unusually dry and lacking in volatile chemicals.

Rivers, lakes, and flood plains

Mars today is generally cold and dry, but there's abundant evidence that it was very different in the past, and much of the landscape is covered with ancient river valleys, flood channels, and lakebeds, while Martian geology also records persistent surface water in the past.

The first signs of Martian rivers, valleys, and canyons were found by NASA's Mariner 9 orbiter in the 1970s. Although the enormous Valles Marineris canyon system ultimately proved not to have been formed by water erosion, elsewhere Mariner found winding, sinuous valleys, often fed by branching tributary networks, that bore the unmistakable signs of ancient flowing water. Since then, countless similar valleys have been identified on the surface. Various alternative explanations for them have been put forward, including wind action and the suggestion that even if the Martian river valleys were carved by water, they could be the result of sudden and catastrophic floods rather than long-running rivers. However, features such as meanders and deep, narrow channels cut into the middle of broader, shallower valleys are both thought to be evidence of rivers that flowed for fairly long periods of time.

Some of the most surprising evidence for ancient rivers, however, comes in the form of "inverted relief"—patterns of winding ridges that appear eerily similar to river valleys and deltas, but are inverted so that they rise above the surrounding terrain. In places, the ridges can be up to 33 feet (10 meters) high and many miles long—they are thought to mark locations in which sedimentary rock formed on the bed of an ancient flowing river, hardened in the period after the river dried up, and has subsequently proved more resistant to erosion by wind and other forces than its surroundings.

Sedimentary rocks offer another line of geological evidence for ancient water—built up by the deposition and compression of layer upon layer of material eroded out of rock elsewhere, on Earth they are usually formed on seabeds, and in lakes and rivers Other mechanisms for creating sedimentary rocks (such as wind deposition) are plausible, but in some parts of Mars—for example, Becquerel crater and the Arabia Terra region (see pages 108 and 114)—these rocks are so complex and abundant that it is hard to imagine them forming without the presence of water. The case for water-lain sediments has been bolstered in the past few years by Mars rover missions. The robots, and Opportunity in particular (see page 192), have been able to study Martian sedimentary rock in situ for the first time, detecting physical features and chemical compounds within them that, on Earth, would be taken as firm evidence of long-standing water.

If the Martian environment was once capable of sustaining liquid water on the surface for long periods of time, then we would expect its rivers to empty into low-lying areas, forming lakes and seas. Mars certainly has features that look like lakes—craters such as Gusev (see page 188), the landing site for NASA's Spirit rover, have shallow floors that suggest they have been filled in by deep layers of sediment. Channels flowing into such craters, and even larger depressions such as Hellas basin (see page 100), show signs of branching river deltas where they once entered large bodies of water. Unfortunately, in Gusev at least, the sedimentary bedrock proved to be buried below a deep layer of mostly volcanic dust and soil that prevented Spirit from investigating it fully.

Perhaps the most dramatic water-related features on Mars, however, are the huge teardrop-shaped islands found in many locations, but most obviously in the Chryse Planitia region (see page 126). These raised, streamlined areas standing amid deeply carved, broad channels are powerful evidence for catastrophic flooding at various times that washed away much of the surrounding landscape. Based on the way such features relate to other elements of the Martian terrain, most experts believe the floods occurred during the Hesperian period, roughly 3.0 billion to 3.7 billion years ago, at a time when exposed surface water was becoming more scarce. Despite the major reservoirs retreating underground, they would still occasionally burst to the surface with disastrous results.

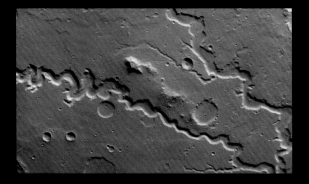

← The sinuous channels of Nanedi Valles bear a strong resemblance to those of terrestrial river valleys, and seem to indicate the presence of water flowing across the Martian surface for a long period of time. Deeper individual channels within the valley floors are also widely viewed as evidence of water action.
→ At first glance this extraordinary network of inverted channels in the southern-hemisphere crater Kasimov seems bizarre, almost artificial in nature. But in reality these branching ridges are the remains of ancient streambeds and gullies, compressed into sedimentary rock that has ended up stronger than its surroundings, and therefore been better able to resist subsequent erosion.

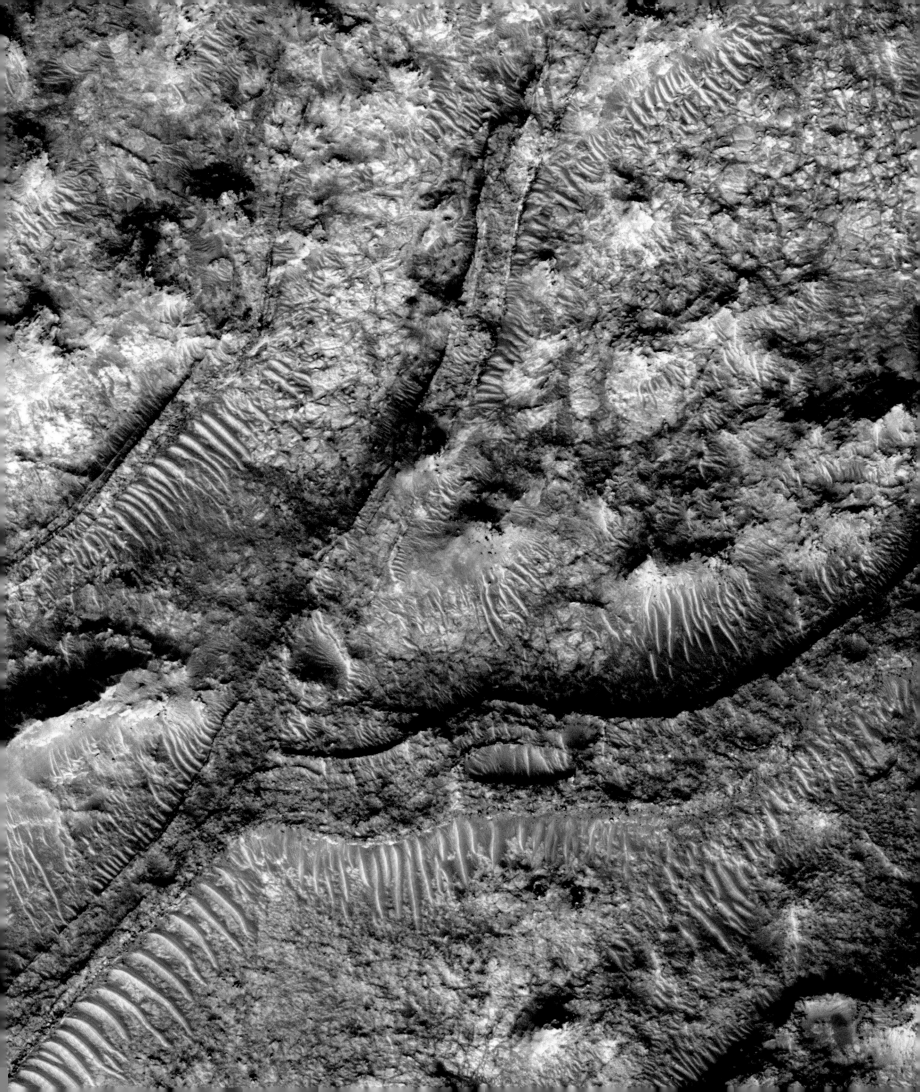

Ancient oceans?

If water once ran freely on the Martian surface, as the evidence of rivers, lakes, sedimentary rocks, and flood plains seems to suggest, then the big question that remains is—how much? According to one theory, the answer is that it covered one-third of the planet to considerable depth.

Supporters of this "Mars ocean hypothesis" claim that the entire extent of the planet's great northern plain, the Vastitas Borealis, is nothing less than the floor of an ancient sea, the Oceanus Borealis. Countless drainage channels seem to suggest that water flowed from the southern highlands into this region, while features that resemble those seen on Earth's shorelines are found along its southern edge, at levels consistent with a single body of water sinking to the lowest possible point.

Perhaps the most convincing of these is a network of inverted relief features (see page 54) that record the path of an ancient river spreading out like a fan as it entered the Aeolis Dorsa region of the northern lowlands. Rivers on Earth spread out in a similar way to form estuaries as they approach large bodies of standing water, and similar features have previously been found on Mars emptying into craters that might once have held large confined lakes. The Aeolis Dorsa "estuary," however, empties straight onto open plains, suggesting the presence of an unconfined, and far more extensive, sea. If such a body of water filled Aeolis Dorsa alone, it could have covered an impressive 39,000 square miles (100,000 square kilometers), and depending on the water level, this might have been just one part of a wider northern ocean.

Supporting evidence for the Oceanus Borealis has come from a new analysis of topographic maps from various space probes, which found that Mars has many more river valleys and channels than previously suspected. The distribution of the valleys in a broad band around the equator and low southern latitudes suggests that they could well have formed as a result of precipitation on the edges of the Oceanus Borealis. Weather systems would have been driven by evaporation from the great northern ocean, but rainfall would not have penetrated far into the highlands, thus explaining the rarity of valleys at high southern latitudes.

Further corroboration has come from the MARSIS instrument on ESA's Mars Express space probe (see page 182). Measuring the properties of the northern plains using radar, it found they have a low density, suggesting they are either composed of relatively lightweight sedimentary rocks or a rock-ice mixture —either of which is compatible with the ocean theory. MARSIS was also able to rule out the idea that the plains are simply unmodified volcanic lavas from when the Martian surface first solidified.

If the ocean hypothesis is correct, then where exactly did the Martian ocean go? Evidence from NASA's Mars Odyssey space probe, which in 2002 found evidence for vast amounts of ice locked up in the soil at high latitudes, suggests that much of it simply froze, mixing with copious amounts of Martian dust that built up in thick sediments over time, to eventually form a thick permafrost. Alternatively, water could have drained into buried aquifers, where it remained in its liquid state for some time before also succumbing to the thinning of the Martian atmosphere and the planet's gradual cooling. Recent discoveries of ragged crater chains running across the Martian plains may be a result of icebergs bumping along the shallow seabeds during colder phases of the planet's past.

Evaporation could also have contributed to the ocean's depletion, with much of the water "boiling" into the air due to gradually falling atmospheric pressure. Clearly, however, today's Martian atmosphere has far less water vapor than Earth's, but this can be explained by a process called atmospheric sputtering, which may have driven many atoms and molecules away into space (see page 26).

↑ → At its greatest extent, the Oceanus Borealis is thought to have had a volume of 14 million cubic miles (60 million cubic kilometers), comparable to Earth's Southern Ocean. The watershed for this vast body of water would have covered more than three-quarters of the planet, and rivers flowing into the ocean would have carried huge amounts of sediment. Over time, these sediments are thought to have accumulated into rock layers up to 330 feet (100 meters) deep overlying the more ancient volcanic bedrock.

Present water?

For much of the 20th century astronomers simply assumed that the red planet would naturally have surface water. Data from the first space probe flybys in the 1960s began to undermine this confidence, and images from Mariner 9 in the early 1970s confirmed the picture of an arid planet.

Today, measurements of the planet's atmosphere and climate have confirmed that large quantities of liquid surface water on Mars are (for the moment at least) an impossibility. In order for liquid water to exist in any environment, the temperature and the pressure must both exceed its so-called "triple point"—and while surface temperatures can indeed rise above the freezing point of water, Martian atmospheric pressure is simply not enough to keep water in liquid form. In most Martian situations, then, water ice heated above freezing point sublimates directly into water vapor, while any liquid water escaping, for example, from subterranean watercourses or aquifers, simply boils on exposure to the thin air. Only a few low-lying areas, most notably the deep Hellas basin in the southern highlands, have sufficient atmospheric pressure for liquid water to linger on the surface, but as yet there is no evidence for it doing so in the recent past. Nevertheless, there are increasing signs that liquid water survives beneath the surface and can occasionally escape.

One convincing line of evidence comes from features known as "gullies," typically found along the edges of steep slopes—for instance, in crater walls, on the sides of old erosion channels formed by ancient water flows or even on dunes, poleward of 30° latitudes. The first of these gullies were found by Mars Global Surveyor in the Gorgonum Chaos region (see page 148), and were assumed to be young features largely because of their pristine appearance—in general, they show very few, if any, impact craters on their surface, implying very recent formation.

Gullies typically consist of numerous small rivulets emerging high up on a slope (in a structure called an alcove), joining together to form a single channel and discharging into a fan-shaped apron at the base. Often, several gullies are found with alcoves at the same level in a crater wall, strongly suggesting that the fluid material that runs downslope emerges from a layer in the Martian soil—either a water-bearing aquifer or a seam of snow- or ice-rich material that melts where it is exposed at the surface. In either case, the flowing material is likely to be water (perhaps with its melting point lowered by substantial amounts of dissolved salt), since it has clearly persisted on the ground for some time. Carbon dioxide, the other obvious alternative, would evaporate far more rapidly on exposure to the thin atmosphere. In 2006, NASA's Mars Reconnaissance Orbiter (MRO) revisited some

sites studied by Mars Global Surveyor five years previously, and found that new gullies had formed where none had been seen before.

But recent though gullies may be, they are still not quite the same as imaging water that is currently present on the Martian surface. The most likely features to show actual flowing water are called seasonal flows or "recurrent slope lineae," and have only been discovered in the past few years by MRO. They are dark streaks that emerge from slopes such as crater walls, growing in extent during the warm seasons and fading during winter. Typically just a few feet wide at most, they can extend for hundreds of feet, and can appear in large numbers. The temperature ranges where they occur would seem to suggest that, like the gullies, they are probably associated with brines—water roughly similar in salt content to Earth's oceans.

Despite appearances, however, these features are not dark because they are actually wet—MRO's CRISM spectrometer (see page 196) has so far failed to detect the presence of water within the flows themselves—the liquid water within them has either evaporated into the air, or remains buried beneath the surface. Nevertheless, geologists have modeled ways in which the flows of brine could disturb the surface sand and darken its appearance, but explaining how the surface brightens again in the winter cold remains a challenge.

→ The long gullies shown in this false-color image from Mars Reconnaissance Orbiter's HiRISE camera are running downslope from the wall of a crater in the southern highlands. The elongated, meandering nature of the gullies and complex features known as "braided channels" within them strongly suggest that they were formed by water flow, while their freshness shows they are the result of relatively recent activity.

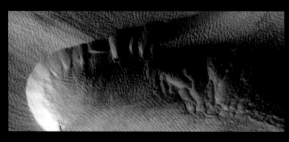

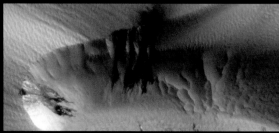

← In contrast to the longer and more persistent gullies attributed to water, these gullies on a dune in the northern Vastitas Borealis are thought to have formed as a result of the sublimation of carbon dioxide ice in Martian spring. The dark patches around these gullies are finer-grained dust pushed to the surface by the escape of buried gas, and these gullies are thought to be landslips caused by the same surface instability.

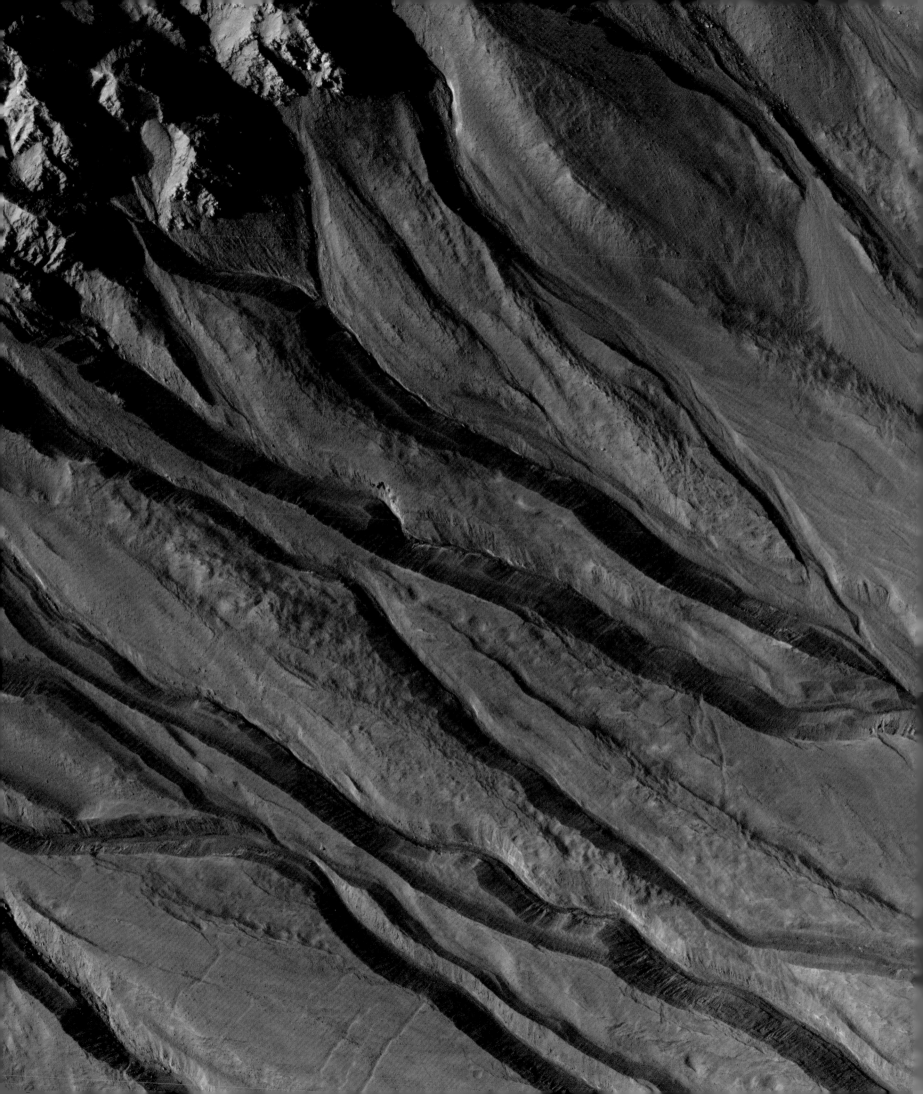

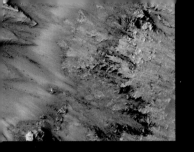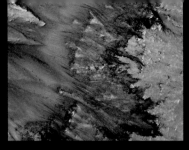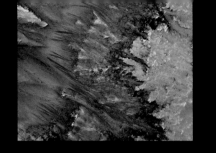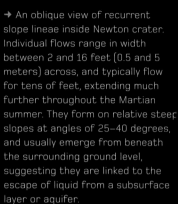

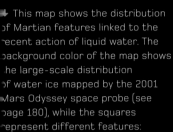

↳ This map shows the distribution of Martian features linked to the recent action of liquid water. The background color of the map shows the large-scale distribution of water ice mapped by the 2001 Mars Odyssey space probe (see page 180), while the squares represent different features:

◼ Warm seasonal flows linked to the flow of briny water.

◼ Surface salt deposits concentrated by water evaporation.

◼ Recent craters exposing fresh water ice.

↑ A sequence of images from Mars Reconnaissance Orbiter shows the evolution of seasonal flows within a large southern-hemisphere crater called Newton. Over the course of a Martian year, the dark features known as recurring slope lineae emerge from higher areas and grow outward and downward in roughly parallel channels, eventually fading at the onset of winter. The lineae do not appear to be water features themselves, but are thought to be created by the action of water flowing just beneath the surface.

→ An oblique view of recurrent slope lineae inside Newton crater. Individual flows range in width between 2 and 16 feet (0.5 and 5 meters) across, and typically flow for tens of feet, extending much further throughout the Martian summer. They form on relative steep slopes at angles of 25–40 degrees, and usually emerge from beneath the surrounding ground level, suggesting they are linked to the escape of liquid from a subsurface layer or aquifer.

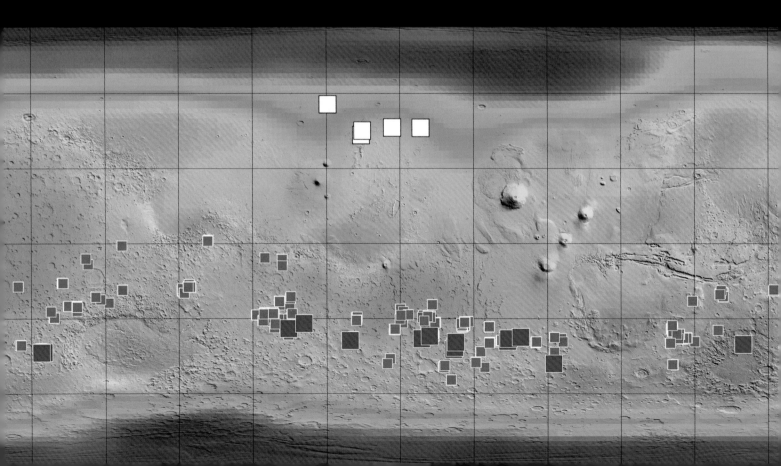

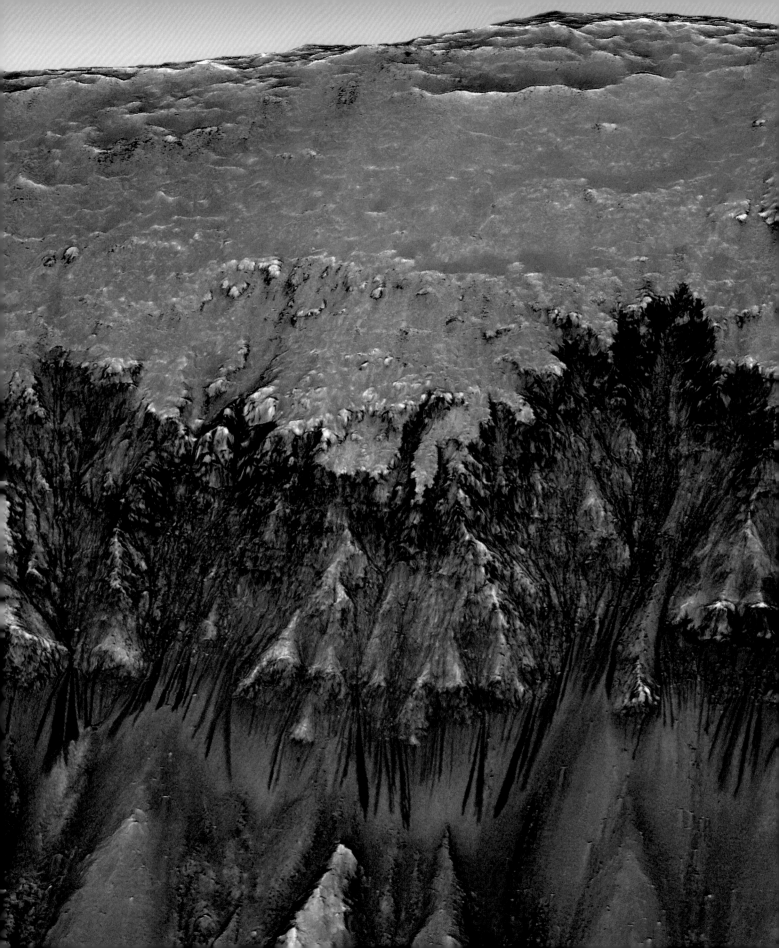

Atmosphere and weather

Despite its sparse nature, the Martian atmosphere still plays a vital role in regulating the Martian environment, transporting heat around the planet and gases from pole to pole, and even generating complex seasonal cloud patterns.

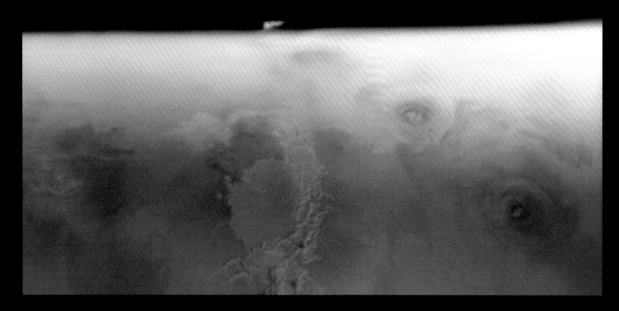

← This Mars Global Surveyor image of the edge or limb of the Martian disk shows the volcano chain of Tharsis Montes (see page 90) disappearing into the atmospheric haze. The distinctive bright feature visible in the upper atmosphere is a so-called "orographic" cloud of dust and water ice that forms above Arsia Mons around the onset of southern winter.

The shell of gases that surrounds Mars and creates its atmosphere is thin and sparse, exerting just 0.6 percent, on average, of Earth's atmospheric pressure at the surface. What's more, there is little of the free nitrogen or oxygen that dominate Earth's air—instead a single chemical compound, carbon dioxide, makes up 95.9 percent of the atmosphere. The inert gas argon is the second most common element, accounting for 2 percent of the atmosphere, with nitrogen accounting for a further 1.9 percent, and oxygen present only in trace amounts. As might be expected of a generally dry, cold planet, water vapor is generally far less common than on Earth, but it does come and go with the seasons, and can form water-ice clouds and even occasional snow. Mystery, meanwhile, surrounds the question of atmospheric methane, confidently reported by several observing teams around 2004, but entirely absent in measurements taken by the Curiosity rover in 2013. Methane is a relatively short-lived gas that should break down quite rapidly in the atmospheric conditions on Mars, and if its presence could be confirmed, it would imply some ongoing mechanism working to generate it—most likely either volcanic activity or, even more intriguingly, some form of life.

The thin Martian air has major effects for the planet's climate: despite the fact that carbon dioxide is a strong "greenhouse gas," it can do little to maintain ground temperatures in such a sparse atmosphere, so the Martian surface tends to rapidly radiate heat and cool down at night, and warm up almost as quickly after sunrise. Vertically,

the atmosphere can be broadly divided into a lower, relatively warm troposphere up to around 19 miles (30 kilometers), a cool mesosphere up to 62 miles (100 kilometers), and a very hot but sparse thermosphere. Above all these, from about 124 miles (200 kilometers) upward, lies the exosphere, a region of trace gases fading away into interplanetary space.

On Earth, atmospheric circulation tends to transport air from warm equatorial regions toward a planet's cooler poles, with air churning over within each hemisphere through three enormous atmospheric cells. In the case of Mars, however, the planet's low "thermal inertia" tends to inhibit this process beyond the region closest to the equator. Instead, other processes tend to dominate and complicate models of the Martian atmosphere. One of these is that, thanks to its low thermal inertia, the atmosphere tends to expand significantly on the sunlit side of the planet, creating "thermal tides" that draw air away from the dark side. Similar winds affect the regional weather, since dark areas of the surface (exposed bedrock with lower brightness or "albedo") absorb heat and warm up more rapidly at sunrise, while brighter areas overlain by loose dust are slower to heat up. Warm, low-pressure areas therefore develop over the dark exposed rock, and create winds as they draw in air from the cooler, high-pressure regions associated with lighter areas. A similar process drives the creation of sea breezes on Earth.

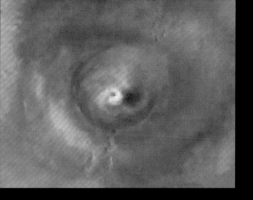 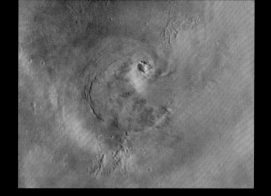 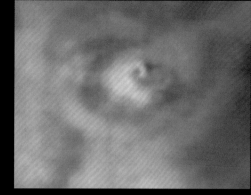

⬆ A sequence of images captured by Mars Global Surveyor during 2001 details the development of the periodic Arsia Mons cloud feature. As sunlight warms the mountain slopes, warm air rises to the peak, where it forms a spiraling column over the caldera trapping dust within it. This in turn seems to trigger the condensation of water-ice clouds.

Another complicating factor is that the planet's relatively fast rotation gives rise to coriolis forces similar to those found on Earth. These create easterly winds (blowing from the east and toward the west) in the tropical regions, and generally deflect the flow of winds clockwise in the northern hemisphere and counterclockwise in the southern hemisphere. Around the poles, long-term interaction between dust, wind, and ice has given rise to spectacular spiral canyons (see page 40).

One other major influence on the Martian climate arises from its delicately balanced temperature. Although conditions today are generally cold enough to keep surface water ice in its frozen form, carbon dioxide is considerably more volatile than water, with a freezing point of −108.4°F (−78°C). What's more, in most conditions it tends to sublimate, transforming directly from its solid form directly into a gas. As the Martian poles go through their seasonal cycle, temperatures at the summer pole typically rise to well above −108.4°F (−78°C), resulting in a large-scale sublimation of ice directly into the atmosphere. At the opposite pole, meanwhile, plunging temperatures draw gas out of the atmosphere and lay it down as thin layers of fresh ice. The end result of this cycle is a pattern of seasonal winds from pole to pole that may reach up to 250 miles per hour (400 kilometers per hour).

Stratosphere
The upper atmosphere extends to altitudes of 62 miles (100 kilometers) or more. Most meteors burn up at around this altitude.

Carbon dioxide ice clouds
Atmospheric CO_2 forms ice crystals at altitudes of around 31 miles (50 kilometers).

Water-ice clouds
Clouds of frozen water vapor form at around 12 miles (20 kilometers).

Dust storm
The largest dust storms can loft dust up to 25 miles (40 kilometers) into the sky.

Seasons of Mars

The combination of an axial tilt similar to Earth's, with an elliptical orbit far more eccentric than any other planet except for Mercury, gives Mars a unique and complex cycle of seasons that affect the higher latitudes of each hemisphere in quite different ways.

The dominant factor behind the seasons on any planet is the varying amount of sunlight received by different hemispheres throughout the year. A few planets, such as Jupiter and Mercury, sit more or less "bolt upright" in their orbits, with their axis of rotation perpendicular to the plane of the orbit. Most, however, are tilted by quite significant amounts—23.4° in the case of Earth, and 25.2° in the case of Mars. A planet's axis points in the same direction in space even as it circles the Sun, so at certain times the northern hemisphere is tipped toward the Sun, and at other times it is tipped away. As both northern and southern hemispheres receive varying amounts of sunlight through the year, they experience seasons: at northern midsummer the north pole is pointing in the direction of the Sun and the south pole points away, while at northern midwinter the situation is reversed. Midway between the two extremes (known as solstices), the two hemispheres receive equal amounts of sunlight at the spring and summer equinoxes.

This much is familiar from Earth, but the eccentric orbit of Mars, which carries it from 128.4 million to 154.8 million miles from the Sun (206.7 million out to 249.2 million kilometers), complicates matters for the Red Planet. Currently, Mars is near perihelion (closest to the Sun) at southern midsummer, and near aphelion (furthest away) at southern midwinter. As a result, the southern hemisphere experiences more extreme winters and summers. In the northern hemisphere, meanwhile, the orbital eccentricity works against the cycle of seasons and tends to produce a more equable climate, with weaker sunlight during the summer months and stronger heating during the winter. What's more, the eccentricity of the orbit means that the length of the seasons (as defined by the time between solstices and equinoxes) varies substantially—northern spring (southern fall) lasts for 193.3 Martian days or "sols," and northern summer (southern winter) for 178.6 sols. Southern spring (northern fall) then lasts for just 142.7 sols, and southern summer (northern winter) for exactly 154 sols. On balance, the result is that northern winters are shorter and warmer than southern ones and the average temperature of the northern hemisphere of Mars remains considerably warmer than that of the southern hemisphere.

The most significant visible effect of the changing seasons is the growth and shrinkage of the polar caps as volatile carbon dioxide ice sublimates from the spring pole and condenses onto the fall one. With each annual cycle, the deposited ice carries dust particles with it, which remain as a so-called "lag" after the ice itself has sublimated once again.

Mars also shows several repeating seasonal weather patterns, the most famous of which are the dust storms that sweep across the planet during perihelion. These frequently originate in the southern tropics, where the effects of solar heating are at their most extreme during the southern summer (see page 64). Around aphelion, meanwhile, clouds of water ice form in the atmosphere during northern summer. Most prominent among these is a "northern annular cloud" that superficially resembles a huge cyclone and forms on summer mornings close to the north polar ice cap, evaporating each afternoon. Other water-ice clouds tend to be seen at this time as a belt around the planet's equatorial region, developing when conditions are just right to allow the condensation of atmospheric water vapor into ice crystals and clouds.

↓ This Mars Reconnaissance Orbiter image captures the first steps in the sublimation of seasonal carbon dioxide frosts trapped within a high-latitude crater at the beginning of Martian spring. The starlike patterns of inverted relief that cover much of the raised mesa in the top half of the crater also appear to underlie the lower, frosted area, and may have formed through countless cycles of ice deposition and sublimation.

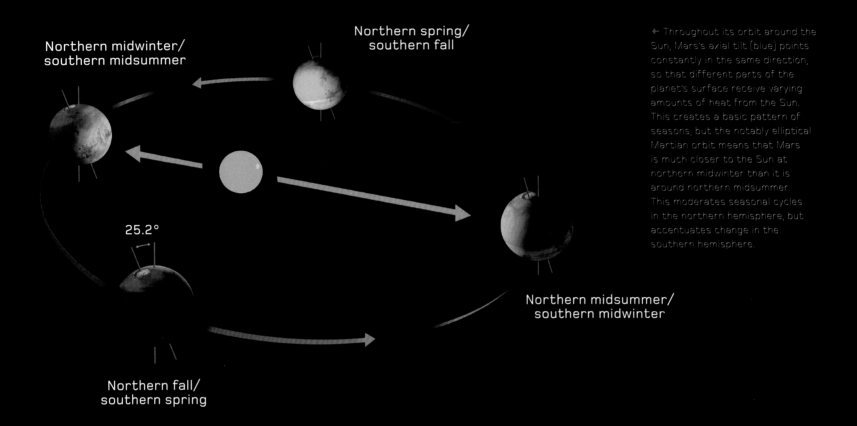

Northern midwinter/
southern midsummer

Northern spring/
southern fall

25.2°

Northern midsummer/
southern midwinter

Northern fall/
southern spring

← Throughout its orbit around the Sun, Mars's axial tilt (blue) points constantly in the same direction, so that different parts of the planet's surface receive varying amounts of heat from the Sun. This creates a basic pattern of seasons, but the notably elliptical Martian orbit means that Mars is much closer to the Sun at northern midwinter than it is around northern midsummer. This moderates seasonal cycles in the northern hemisphere, but accentuates change in the southern hemisphere.

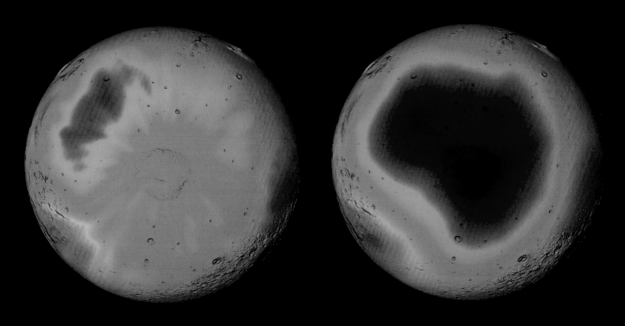

← This pair of images from NASA's 2001 Mars Odyssey mission trace the distribution of water ice at the surface of the northern hemisphere during winter (left) and summer (right). During winter, most of the water ice is hidden beneath thick layers of carbon dioxide frost, while in summer it is exposed at the surface, both as fresh ice or locked within the permafrost.

Martian climate change

One of the most intriguing questions about Mars is just how different it was in the past, and whether it might change again in the future. There's plenty of evidence for water on the surface in the distant past, but do conditions on Mars alter over shorter timescales?

Some of the most important evidence for this idea comes from research into Earth's own prehistoric ice ages. Our planet has gone through a number of major ice ages over the 600 million years of relatively "recent" history, and the fossil and geological records suggest that these are mainly linked to long-term changes in geography (as the continents move around on Earth's surface) and equally dramatic changes in the composition of the atmosphere. These major ice ages may each last for millions of years, but within them, geologists have identified much shorter-term fluctuations between cold periods, known as "glacials," and warmer intervals, or "interglacials." Technically speaking, the current warm period within which human civilization has evolved and flourished over the past 10,000 years is merely the latest interglacial in an ice age that may still be ongoing.

It now seems that relatively short-term cycles of glacials and interglacials are due to minor changes in Earth's orbital parameters. Known as Milankovitch cycles, after Milutin Milankovitch, the Serbian scientist and mathematician who discovered them during the 1920s, there are three distinct patterns: a 25,800-year "wobble" in the direction

of Earth's axial tilt, known as the precession of the equinoxes, which affects how far the Sun is from Earth at different points in the seasons; a slight 41,000-year variation in the angle of Earth's axial tilt (and hence the severity of the seasons); and a 413,000-year "flexing" in the shape of Earth's orbit between near-circular and more elongated with greater differences in the amount of heat reaching Earth throughout the year. Together, these cycles interweave, canceling out each other's influence at some times and reinforcing it at others, to create a complex pattern of glacials and interglacials.

We could reasonably expect similar cycles to affect Mars, but the fact that they are so much more severe might come as a surprise. We've already seen how Mars's highly elliptical orbit produces less extreme temperatures in the northern hemisphere throughout each year and greater variations in the southern hemisphere (see page 64), but Mars's 171,000-year precession of the equinoxes reverses this situation regularly. The Martian orbit, on the other hand, is currently about as elliptical as it can get—at other times in its 100,000-year cycle it can become almost circular.

↓ A stunning image from the Mars Reconnaissance Orbiter's HiRISE camera homes in on a region of the south polar ice cap to reveal interconnected flat-bottomed pits whose golden walls exposed the layered deposits just beneath the pristine surface. Elsewhere on the ice cap, repeated imaging of similar pits has revealed their steady year-on-year expansion, suggesting that Mars is currently going through a sustained period of global warming.

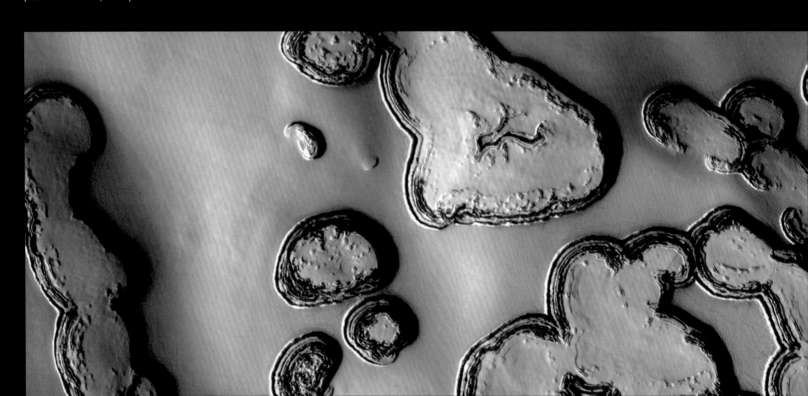

Most dramatic of all, though, is the 124,000-year variation in the planet's axial tilt, which currently ranges between about 15° and 35°, but can show even more extreme variation over millions of years. Changes to the shape of the Martian orbit are driven in part by its relative proximity to the giant planet Jupiter, and the influence of other planets also drives unpredictable long-term changes in axial tilt. Mars is particularly vulnerable to these factors because it has no large satellite, equivalent to Earth's Moon, to help stabilize it.

Although the way in which Martian Milankovitch cycles interact is still poorly understood, the general scientific consensus is that Mars is currently in the interglacial phase of a much longer ice age that may have persisted through the past 5 million years. Using images of the south polar cap taken by various orbiting satellites, scientists have tracked the expansion and erosion of so-called "Swiss cheese" terrain (holes in the ice, created by seasonal loss of ice) and found that the erosion appears to be on the increase from year to Martian year. This might suggest that the planet is currently undergoing an increase in average temperature analogous to (though with different causes from) Earth's own recent climate change.

Martian ice ages are nothing like those of Earth's recent past, however—they seem to coincide with periods when the planet's axial tilt is near its least extreme, resulting in *warmer* average temperatures. Water freed from its permanent high-latitude locations is transported toward equatorial regions where it can develop into glaciers (see page 44). This is thought to explain many of the glacial features that are found today at comparatively low latitudes, thinly concealed beneath layers of dust.

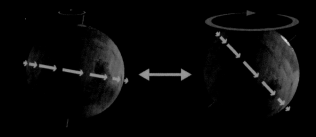

← With no large moons to stabilize its inherent "wobble," the axial tilt of Mars wanders back and forth between around 15° and 35° in a roughly 124,000-year cycle.

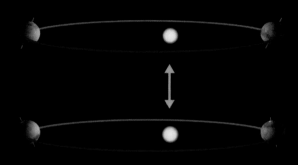

← Precession of the equinoxes sees the Red Planet's axis of rotation change its orientation in a 171,000-year cycle, altering the severity of the seasons experienced in each hemisphere.

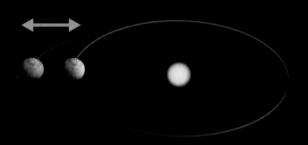

← The influence of other planets (and Jupiter in particular) causes the Martian orbit to flex in shape in a roughly 100,000-year cycle, varying between distinctly elliptical (as at present) and almost circular.

→ This artist's impression depicts Mars as it may have looked during a recent ice age, between around 2.1 million and 400,000 years ago. Around this time, the interlocking Milkankovitch cycles combined to create a period of general polar warming that released more water vapor into the atmosphere and ultimately allowed ice to spread down to latitudes of about 30° north and south of the equator. Much of that ice persists in the Martian soil to the present.

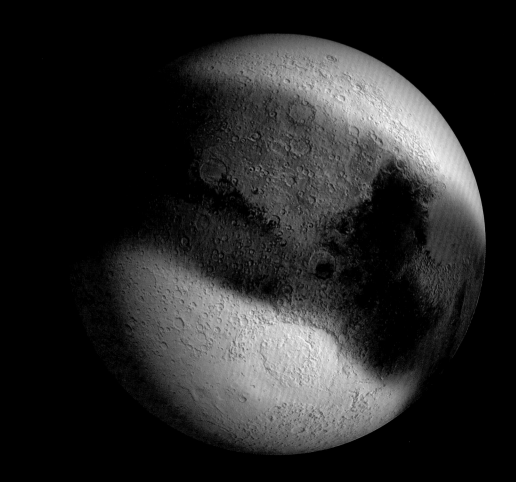

Although some astronomers had speculated about the occasional yellow clouds they saw on the Martian clouds since the 19th century, the existence of Martian dust storms was only confirmed in 1971 when Mariner 9 swung into orbit and found its view of Mars entirely obscured. The ultimate power source driving these storms (and indeed all Martian weather) is the Sun, so it should be no surprise that they typically peak around perihelion, when Mars is at its closest to the Sun and receiving up to 40 percent more solar heating. Storms usually begin in equatorial regions, where the difference between day- and nighttime temperatures, and the associated expansion of the atmosphere and drop in pressure, are at their greatest. In these conditions, light breezes begin to loft dust into the air, where it can hang for a considerable time.

Martian dust storms can persist for longer periods and travel across much longer distances than on Earth, because of the planet's plentiful supply of surface dust and a lack of atmospheric water vapor. On Earth, water vapor causes dust particles to stick together, eventually falling out of the air under their own weight. However, the mechanisms that drive regional storms to explode on a global scale are still poorly understood, though they are probably linked to large-scale temperature changes that may be both a cause and an effect of the storms.

During a major storm, surface temperatures on Mars can drop by several degrees, while atmospheric temperatures can rise by up to 86°F (30°C), as individual dust particles absorb solar radiation before it can heat the ground.

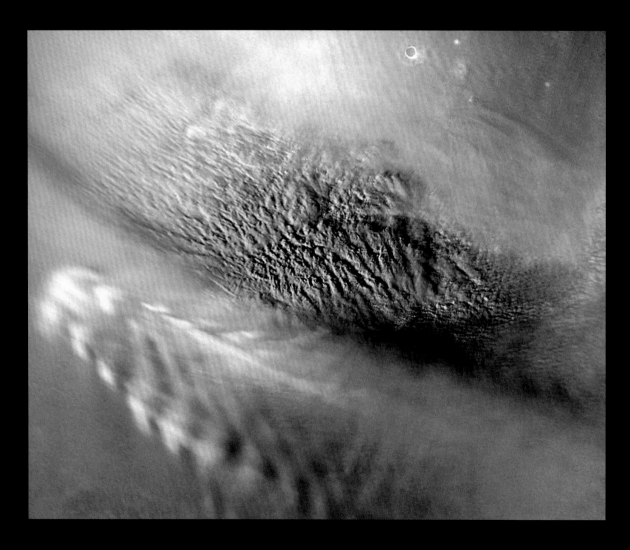

← At the center of this image, an orange-tinted dust storm looks deceptively solid as it sweeps across Utopia Planitia close to its boundary with the seasonal ice cap, near the end of northern winter. Despite its apparent solidity, the storm, captured by Mars Reconnaissance Orbiter in November 2007, flared up and subsided in less than a day. The distinctive water-ice cloud sweeping across the bottom of the picture, meanwhile, is a so-called "lee-wave" cloud, created by the change in temperature and atmospheric pressure over nearby Mie crater.

→ A pair of Hubble Space Telescope images captured around the time of Mars's 2001 opposition (close approach to Earth) chronicles the eruption of the largest global dust storm seen since the early 1970s. In the upper image (photographed in June 2001), the storm's progenitor can be seen already overspilling the edges of the Hellas basin (at about the four o'clock position on the Martian disk). By the time the lower image was taken in September, the storm had engulfed the entire planet for almost two months, leaving only the most prominent features visible through its orange haze.

According to one theory, once warm air around a regional storm accumulates in sufficient amounts or encounters other specific weather conditions, it can rapidly expand out to colder areas of the planet, creating violent winds and sudden drops in pressure that create a "snowball effect."

Wind speeds within the storms can reach up to 100 miles per hour (160 kilometers per hour), though thanks to the rarefied nature of the Martian atmosphere and the small size of the dust particles involved, they are not as physically devastating as their speed alone would imply. Photographs taken from Mars landers and rovers engulfed within storms show little sign of actual dust in motion. Instead, the storms make their influence felt through a gradual buildup of dust on exposed surfaces, which often causes solar panels to experience potentially dangerous drops in power output.

The mechanism that brings the storms to an end is just as poorly understood, though it may be that once a storm becomes truly global, temperature variations are evened out and the driving winds simply die away. Large storms can certainly persist for several weeks, while it can take months for the levels of atmospheric dust to return to normal and surface temperatures to recover.

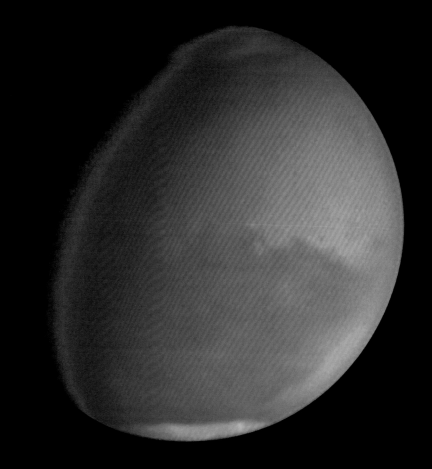

or so. These towering columns of swirling air draw dust up from the surface and expose the darker bedrock beneath in squiggling patterns.

Dust devils occur when hot air near the ground rises suddenly due to the movement of cooler, low-pressure air above it. Because the air generally gets cooler at high altitudes, this process is self-sustaining and the hot air develops into a rising column. As the air column narrows, random movement in the original air mass becomes intensified in accordance with the conservation of angular momentum—just as a pirouetting skater can speed up by pulling her arms in toward her body. Rotation at the base draws in more and more hot air from the surroundings and allows the dust devil to grow and sustain itself. Like a tornado, the swirling mass of air is theoretically transparent, but tends to draw up loose dirt and other material from the ground, which makes it visible.

Martian dust devils can be many times the size of those seen on Earth—hundreds of feet across and many miles high. Unsurprisingly, they tend to occur early in the morning near midsummer in each hemisphere—around the time when the Sun's heating effect on the lower atmosphere is at its strongest. Wind speeds within them can reach 62 miles per hour (100 kilometers per hour) or more, but the low pressure of the Martian air means that, like dust storms (see page 68), they present little danger to spacecraft on the surface. Mars

Pathfinder's weather station detected the passage of several dust devils during 1997 (though it did not photograph them), while the more recent rovers have positively benefited from encounters with dust devils that have helped clear away accumulated dust from their solar panels, rejuvenating their power supplies.

The dark tracks left behind by the passage of these swirling dust columns, meanwhile, were first identified in early photographs from Mars Global Surveyor, and a link to dust devils was soon suggested. However, it was December 1999 before Surveyor delivered an image of a dust devil "caught in the act," scouring the landscape clean to leave a dark trail behind it. Just as on Earth, the geography of certain areas and features on Mars makes them a magnet for this kind of activity, leaving them with the appearance of a child's drawing pad until their surfaces are wiped clean again by the passage of a major dust storm.

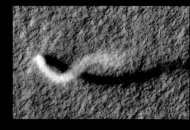

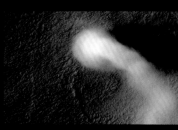

↑ Two images gathered by Mars Reconnaissance Orbiter in early 2012 capture dust devils scouring the northern plains. The upper image is a relatively small whirlwind about 0.5 miles (800 meters) high and 100 feet (30 meters) across, while the lower one is about twice as wide, but much taller, rising around 12 miles (20 kilometers) into the Martian sky.

← Dust devil tracks appear like random scribbles on the dusty plain at the center of Schiaparelli, a 286-mile (461-kilometer) impact basin just south of the Martian equator. Because the exposed darker bedrock beneath the bright dust tends to absorb heat better, dust devils tend to form repeatedly in the same areas.

→ Broad dust devil tracks create chaotic patterns in a dune field to the north of Syrtis Major. Because the exposed darker bedrock beneath the bright dust tends to absorb heat better, dust devils tend to form repeatedly in the same areas.

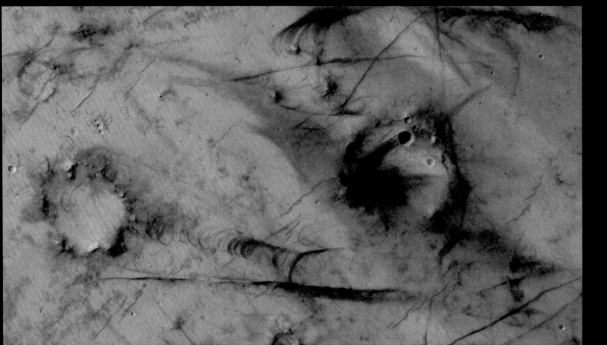

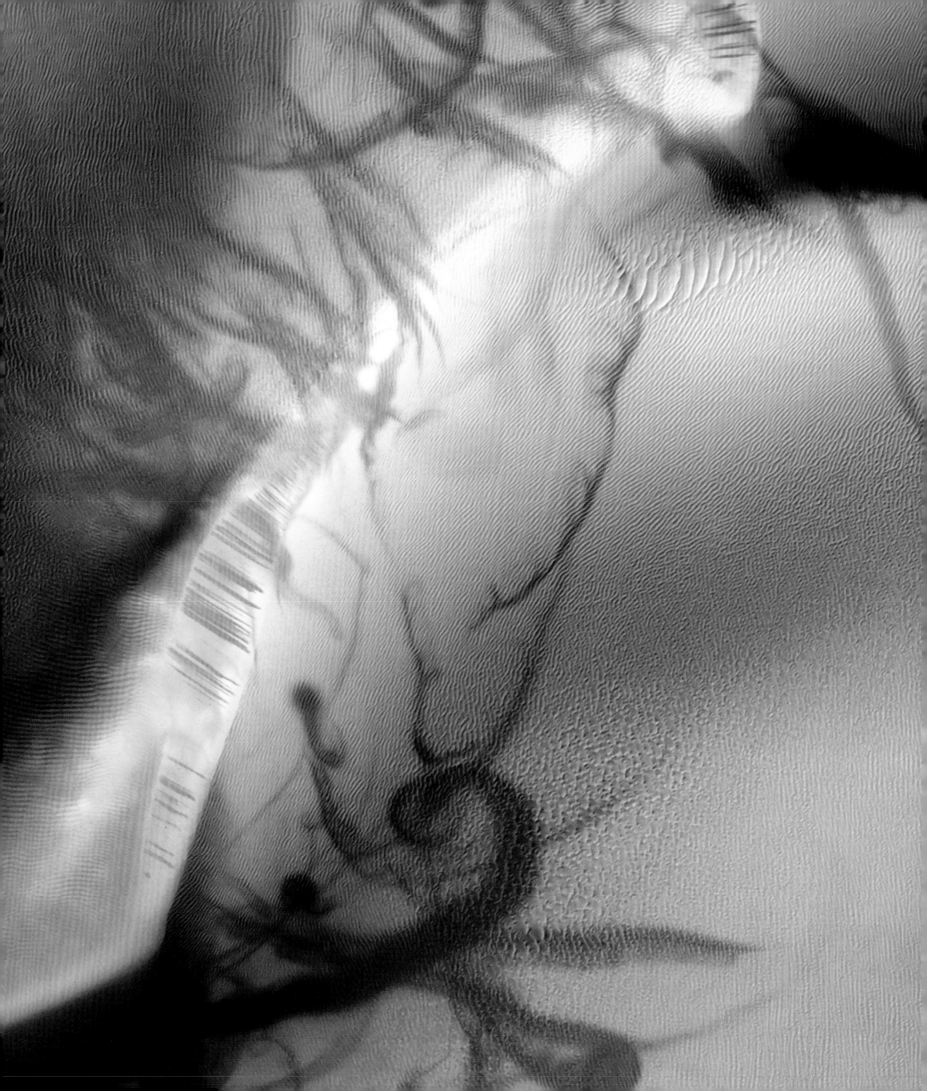

The first space probes to reach Mars in the 1960s dealt a heavy blow to those still hoping that the Red Planet might be a welcoming venue for moderately complex forms of life, although it was clear to most before then that the Martian atmosphere was too thin and dry, with the wrong constituents for Earthlike life. However, the early discovery of features that looked like dried-up riverbeds held out a flicker of hope that Mars had been more hospitable in its distant past, and decades of further research have only strengthened the case.

It's worth noting that the search for conditions similar to Earth in the quest for alien life is not simply a hangover from the geocentricism of earlier ages—there are sound scientific reasons for believing that life in any form would necessarily rely on the same essential chemistry of complex carbon-based "organic" molecules. Furthermore, such molecules would require the presence of a suitable fluid solvent in order to undergo chemical reactions, of which liquid water is by far the most effective.

Meanwhile, as our space probes have been searching for potentially hospitable conditions, past and present, on other worlds in the solar system, biologists have been discovering that life on Earth is far more varied than we once thought, and capable of withstanding much more hostile environments. Locations such as the highly alkaline Mono Lake in California and the acidic Grand Prismatic Spring in Yellowstone, Wyoming, have been shown to play host to "extremophile" microorganisms that function by processing chemicals from their environment in ways we had never previously suspected. Primitive life has been found everywhere from superheated "hot rocks" deep underground, to the cold, dry wastes of Antarctica. The requirement for Earthlike conditions, it seems, may not be as limiting as was once thought.

So, could life have once evolved on Mars—and might it still cling on today? The first serious attempt to answer this question was made by NASA's Viking landers in the 1970s (see page 174). Each of these spacecraft carried out a series of tests to look for possible living organisms or organic remains in the Martian soil. Most produced negative results, but one in particular—the Labeled Release (LR) experiment—was enticingly inconclusive.

This experiment involved feeding a soil sample with nutrients "tagged" with the radioactive isotope carbon 14 and monitoring it for the possible release of radioactive gas (carbon dioxide) that would be

← Mono Lake is a large shallow and highly alkaline soda lake in California, with hot springs and rising towers of calcium carbon known as tufa. Despite its apparent hostility, it is home to an entire ecosystem based on extremophile organisms. Based on evidence for similar hot springs in the distant past of Mars, NASA scientists have use this strange environment to test procedures for finding life on the Red Planet.

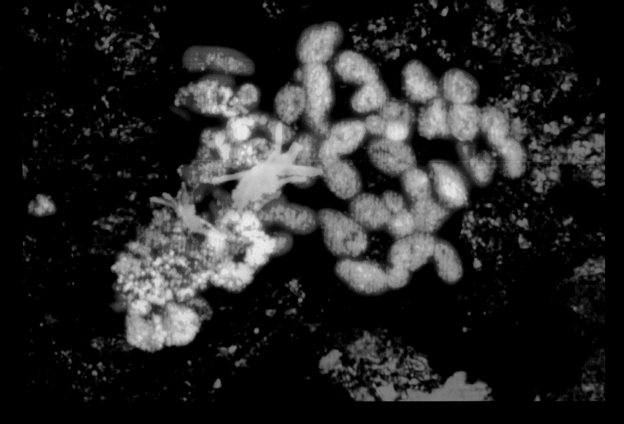

← SLIMEs (subsurface lithoautotrophic microbial ecosystems) are colonies of extremophile bacteria that have been discovered living deep below ground on Earth since the 1990s. Existing without light or oxygen, they are thought to feed on hydrogen from rock and generate methane as a waste product—could similar organisms exist on Mars?

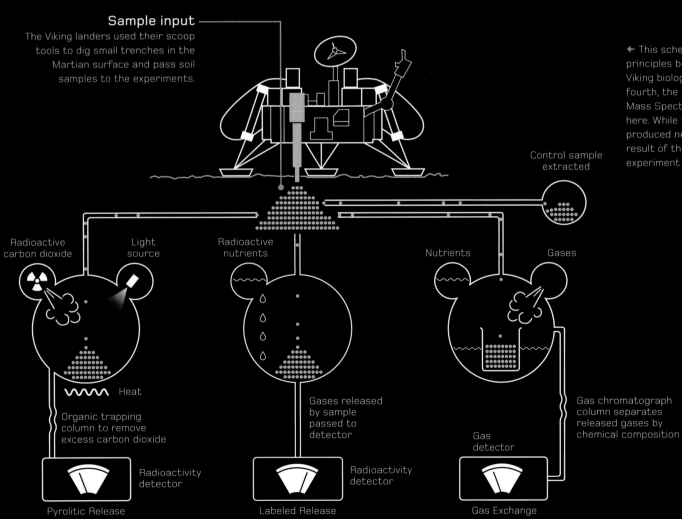

Sample input

The Viking landers used their scoop tools to dig small trenches in the Martian surface and pass soil samples to the experiments.

← This schematic shows the principles behind three of the four Viking biological experiments—a fourth, the Gas Chromatograph— Mass Spectrometer, is not shown here. While three of the experiments produced negative results, the result of the Labeled Release experiment remains controversial.

Control sample extracted

Radioactive carbon dioxide

Light source

Radioactive nutrients

Nutrients

Gases

Heat

Organic trapping column to remove excess carbon dioxide

Gases released by sample passed to detector

Gas chromatograph column separates released gases by chemical composition

Gas detector

Radioactivity detector

Radioactivity detector

Pyrolitic Release

Labeled Release

Gas Exchange

produced by microbes in the soil processing and metabolizing the nutrients. Intriguingly, both Viking LR experiments showed exactly this kind of behavior at first, producing a steady stream of carbon dioxide. But frustratingly, when the experiments were repeated a week later, neither produced any further gas. Arguments about the precise meaning of these results have raged ever since—some have argued that the experiment did indeed find life, but somehow killed it off before the second phase, while others have claimed that the Martian soil might contain highly reactive "superoxide" chemicals capable of producing an initial burst of gas without the need for microbes.

The two-decade gap in the exploration of Mars that followed the Viking missions left many questions unanswered, and the only subsequent mission designed to specifically search for traces of life, the Beagle 2 lander attached to ESA's Mars Express mission, sadly met with failure (see page 182). NASA's orbiters, rovers, and landers, meanwhile, have focused on learning more about the past and present environment, partly in order to understand whether life could ever have got a foothold in the first place. Discoveries such as hematite minerals in the Martian rocks (see page 114) have more or less settled the case for a warmer, wetter Martian past, while Spirit's excavation of silica-rich soil (see page 188) suggests the presence of "hot spring" environments that would be ideally suited for the evolution of life

In 1996, meanwhile, a scientific controversy erupted when NASA scientists claimed to have found fossil evidence for past Martian life. While we have yet to return rock samples directly from the surface of Mars, geologists have come to realize that natural processes occasionally bring Martian rocks directly to their doorstep in the form of meteorites. Blasted free of Martian gravity by asteroid impacts or even major volcanic eruptions, these rocky fragments may spend many millions of years orbiting between Mars and our planet before ultimately crashing to Earth.

Most meteorites, of course, originate among the rocks of the asteroid belt—debris left over from the formation of the solar system itself. Martian meteorites were identified in the early 1980s based on clear differences from the vast majority of their brethren, and similarities to the rocks analyzed by the Viking Mars landers. The controversial meteorite designated as ALH 84001 was discovered in the Allan Hills of Antarctica in 1984, but its remarkable features did not become clear until more than a decade later.

ALH 84001 is thought to have been formed during a large meteorite strike, perhaps in the Eos Chasma region of the Valles Marineris, around 4 billion years ago, but remained on the Martian surface until it was ejected into space by a smaller impact some 15 million years ago, arriving on Earth a mere 13,000 years ago. Studying the rock with a scanning electron microscope, the NASA team discovered tiny wormlike structures just a few tens of nanometers (billionths of a meter) across, which they proposed might be the fossil remnants of tiny "nanobacteria," just a fraction of the size of the smallest microorganisms known on Earth. Just as significantly, alongside the "fossils," the rock carried traces of minerals such as magnetite that would, on Earth, be taken as the work of living organisms.

Needless to say, these claims were greeted with considerable skepticism, and within a few years other scientists (including a second NASA team) showed ways of producing many of the features of ALH 84001 without the need for life, including the wormlike "nanobacteria" structures. The size of the purported nanobacteria, meanwhile, was also problematic, as some researchers questioned whether such tiny organisms could feasibly exist. The original team have since responded by raising doubts about whether the proposed alternative mechanisms for mimicking the various features could work in practice, and identifying so-called "biogenic" structures in other Martian meteorites. Given the uncertainties that inevitably surround much of ALH 84001's history, all that can be said is that the evidence remains inconclusive, and the questions surrounding fossil Martian life are likely to remain unresolved until a future robot or human mission can return pristine Martian rocks to Earth from a thoroughly understood geological context.

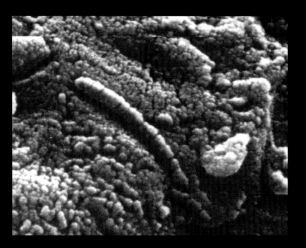

← ← This unassuming lump of rock is the controversial Martian meteorite ALH 84001. Found in Antarctica in 1984, it certainly offers scientists a rare sample of rock from ancient Mars, but does it also preserve fossils of primitive Martian life?
← An electron micrograph shows wormlike structures within the ALH 84001 meteorite. Though reminiscent of terrestrial bacteria, the scale of these nanofossils, 20–100 nanometers in diameter, is far smaller than any known Earth microbes.

The past decade has seen the controversy around Martian fossils supplanted by another mystery —possible signs that Mars is alive, at least in some sense, right now. In 2003, scientists using Earth-based infrared telescopes on the Hawaiian island of Mauna Kea reported the detection of small but significant amounts of methane (a few parts per billion) in the Martian atmosphere, concentrated over specific parts of the surface such as Nili Fossae (see page 160). The following year, the discovery was backed up by ESA's Mars Express probe in orbit around the Red Planet, and further observations over the next few years continued to strengthen the case for varying amounts of the gas.

Methane is significant because it is highly unstable in Martian conditions, and should quickly break down through the action of the Sun's ultraviolet light and other atmospheric gases, in periods of between a few months and a few years. In order for methane to be present, it must therefore be continuously regenerated, and the most likely mechanisms to do that are linked to either volcanic or hydrothermal activity, or the action of "methanogen" bacteria. The confirmation of active present-day volcanism on Mars would be a major advance in our understanding of the planet, while the discovery of bacteria would be revolutionary. Unfortunately, the chances of an imminent breakthrough suffered a blow when NASA's Curiosity rover reported finding no trace of methane in the atmosphere at its landing site, and some skeptics now argue that the most convincing methane observations may simply be picking up traces of the gas in Earth's own air. New spacecraft designed to study the Martian atmosphere in more detail may settle the question once and for all in the next few years, but for now, as with the case of the putative fossils, the jury is still out.

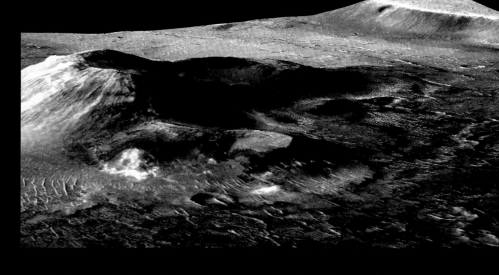

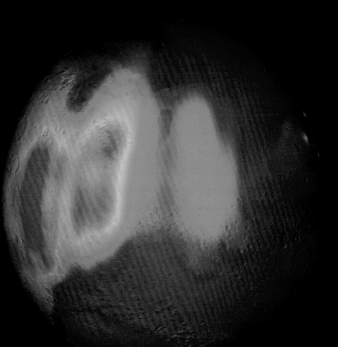

← This view of Mars maps spectroscopic signatures associated with the suspected presence of methane onto a Martian globe. Warmer colors indicating increasing levels of methane seem to be concentrated around the Nili Fossae region (see page 160).

↑↑ This false-color perspective view of the Nili Patera volcanic cone is based on infrared observations and indicates the presence of bright hydrothermal deposits at the foot of the caldera and on one flank. Such hot, steamy environments would have been ideal for microbial life to get a foothold.
↑ The Grand Prismatic Spring in Yellowstone National Park, Wyoming, is the largest hot spring in the United States. It owes its vivid colors to a variety of microbes that grow within and around the 160°F (70°C) spring, and rely on its mineral-rich waters for survival— only the hottest waters upwelling at the center of the spring are sterile. The unexpected ability of microbial life to thrive in seemingly hostile environments has transformed our thinking about the prospects for life on other planets.

Phobos and Deimos

The two Martian moons were discovered within days of each other in August 1877 by astronomer Asaph Hall, working at the US Naval Observatory in Washington D.C., as the result of a deliberate and painstaking search.

Phobos, the inner of the pair, has dimensions of 17 x 14 x 12 miles (27 x 22 x 19 kilometers), while the outer moon Deimos is 10 x 7 x 6 miles (16 x 12 x 10 kilometers). Phobos orbits Mars in just 7 hours 40 minutes, at an average height of just 3,720 miles (5,989 kilometers) above the surface, so that despite its tiny size, it appears fully one-third the diameter of Earth's Moon in Martian skies. What's more, its rapid orbit outpaces the planet's own rotation, so that Phobos appears to rise in the west and set in the east, roughly twice each Martian day.

Deimos's orbit, in contrast, is a slightly more leisurely 30 hours 19 minutes at an average height of 12,466 miles (20,070 kilometers). When combined with the planet's own rotation, this means that it moves much more slowly as seen from the ground, rising in the east and taking around 2.7 days to cross the sky and set in the west.

Phobos and Deimos have been quite well studied during close encounters with various orbiting space probes, but they still have many mysteries. Physically, they look quite different—Phobos is heavily cratered and also scarred by puzzling parallel lines, while Deimos seems less pockmarked by impacts, and those craters which can be seen are softened by what seems to be a layer of dust.

Superficially, both these tiny worlds bear a strong resemblance to asteroids. At first, their dark, cratered surfaces and the spectra of their light seemed to match those of a common class called carbonaceous or C-type asteroids, and for a long time, they were generally assumed to simply be strays from the asteroid belt, captured by Martian gravity as their orbits brought them through a close encounter.

However, recent studies have suggested otherwise, and there is growing support for the idea that both Phobos and Deimos formed in the same way as Earth's Moon—originating as debris flung into orbit around the planet during a major impact on the surface of Mars.

Subject to strong tidal forces, Phobos's orbit is losing a little more height each year, and astronomers estimate that within about 40 million years it will come so close to Mars that it will no longer be stable. Phobos may then either plunge to the ground forming a huge new crater or break up in orbit to form a ring around the Red Planet.

→ This stunning close-up of Phobos was captured by Mars Reconnaissance Orbiter during a close flyby of the Martian moon in 2008. It focuses on an enormous crater called Stickney, some 5.6 miles (9 kilometers) in diameter, which dominates a huge area of the Moon's surface. Stickney is named in honor of Asaph Hall's wife, mathematician and political campaigner Angeline Stickney Hall. The impact that formed it must have come close to shattering the Moon completely.

↓ Two more images from the HiRISE camera aboard Mars Reconnaissance Orbiter show the two Martian moons roughly to scale and highlight the striking differences in appearance between them. Phobos is cratered and scarred while Deimos is brighter and smoother, apparently due to a thick layer of dusty regolith. One theory to explain the difference is that Deimos may be little more than a loose "rubble pile," allowing it to absorb and dissipate the energy of impacts without major disruption to the surrounding surface.

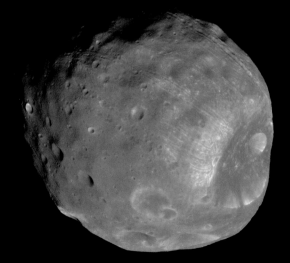

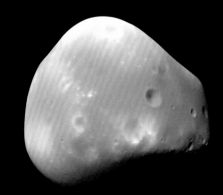

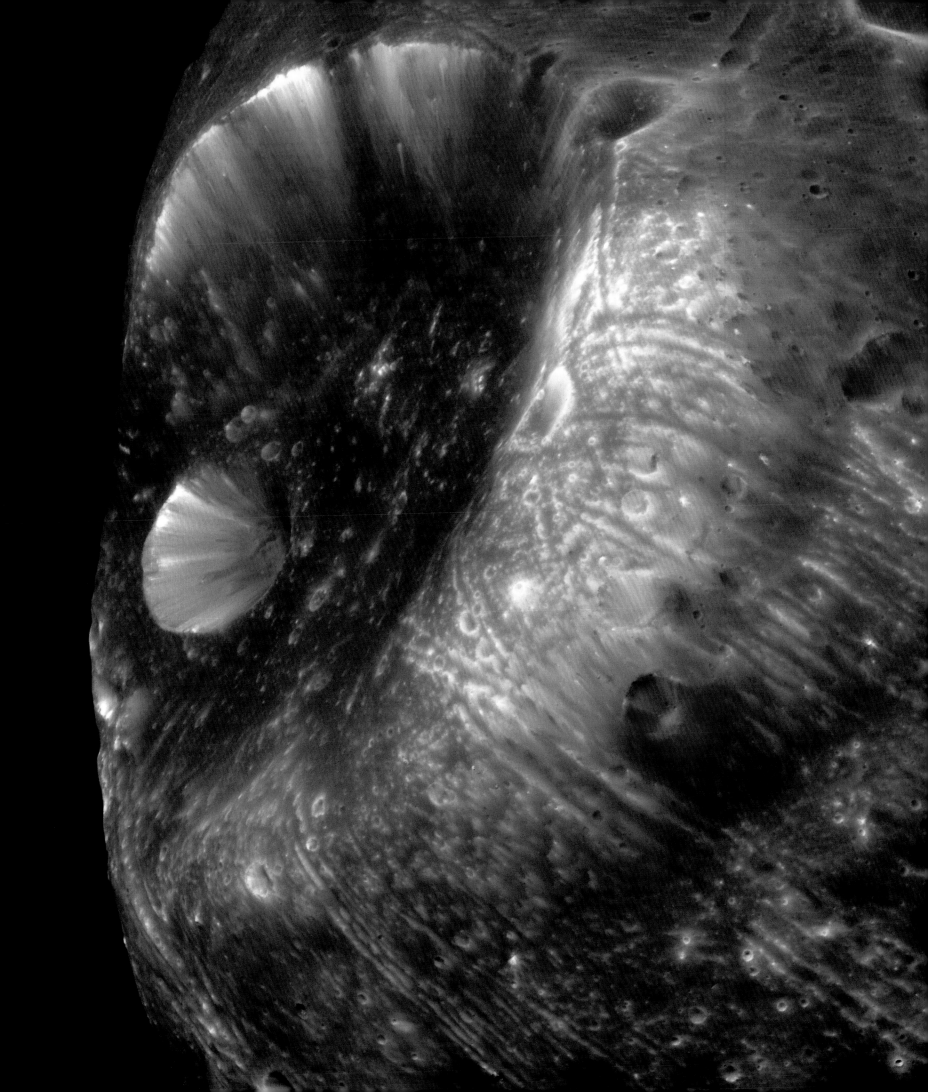

geology and geography gives rise to a stunning variety of landscape features. From cold northern plains, vast dune fields, and towering volcanoes, to plunging canyons, sinuous river valleys, and cratered uplands, the Red Planet reveals many different sides to its personality.

→ A satellite view of the wind-sculpted landscape around Victoria Crater.

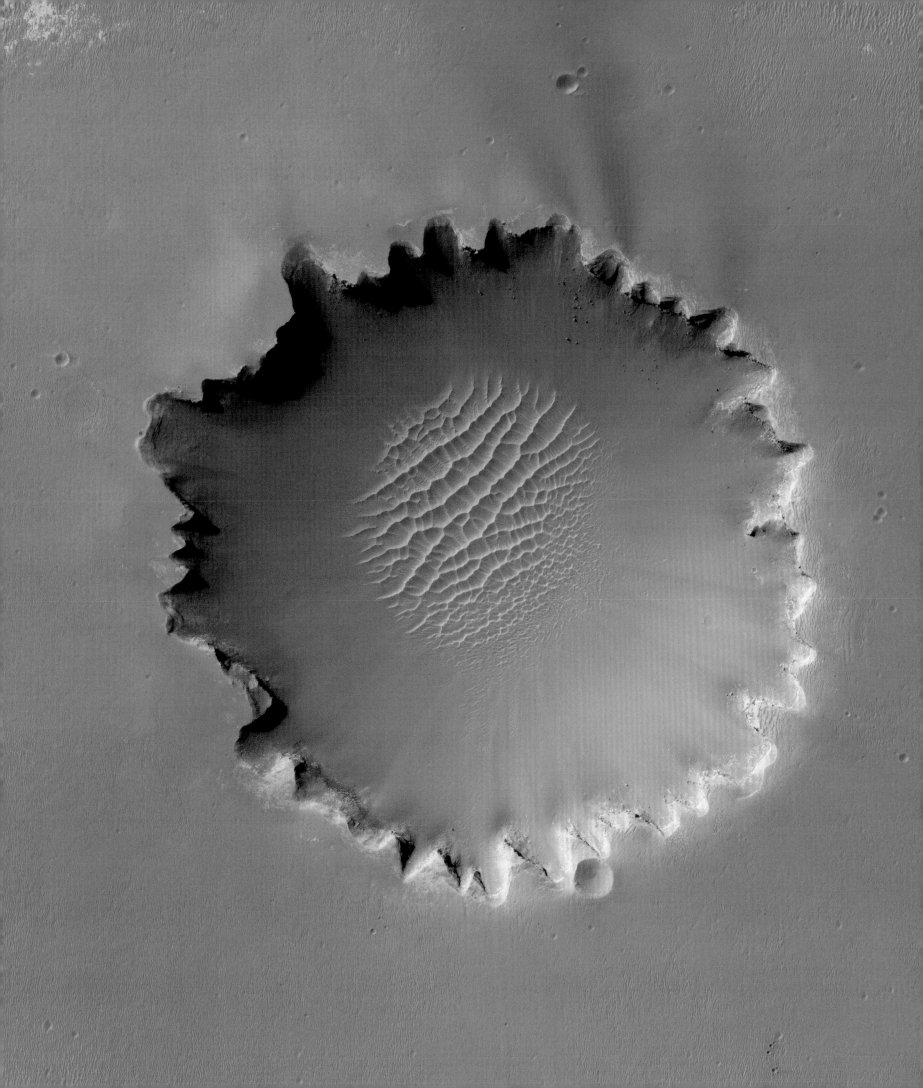

Mapping Mars

The four maps shown on this and the following pages show the major surface features of Martian hemispheres centered on longitudes 0°, 90°E, 180°E, and 270°E (90°W). The Martian prime meridian of 0° longitude is arbitrarily centered on a small southern-hemisphere crater called Airy-0.

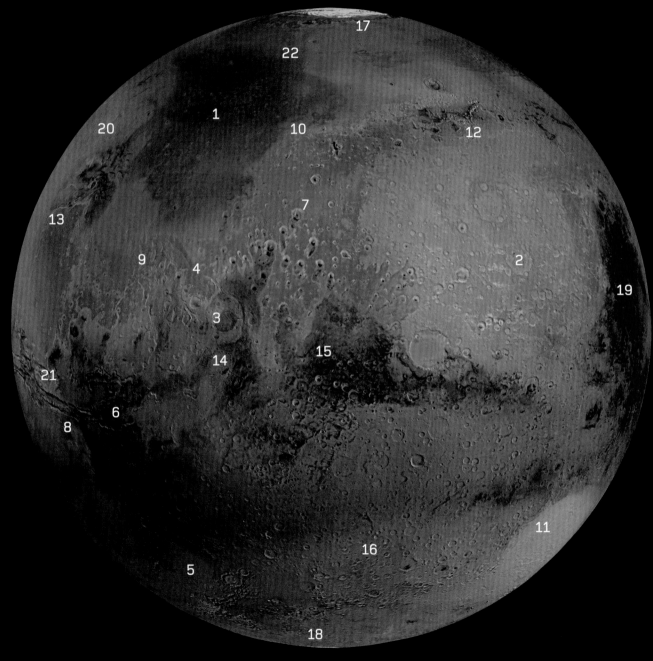

Centered on 0°

0°

1 Acidalia Planitia	**6** Aureum Chaos	**11** Hellas basin	**16** Noachis Terra	**21** Valles Marineris
2 Arabia Terra	**7** Becquerel crater	**12** Ismenius Lacus	**17** North pole	**22** Vastitas Borealis
3 Aram Chaos	**8** Capri Chasma	**13** Kasei Vallis	**18** South pole	
4 Ares Vallis	**9** Chryse Planitia	**14** Margaritifer Terra	**19** Syrtis Major Planum	
5 Argyre Planitia	**10** Cydonia	**15** Meridiani Planum	**20** Tempe Terra	

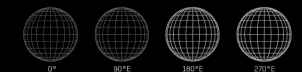

← Throughout this chapter, color coded icons on each gazetteer entry show the approximate location of each site. To find the location on the main maps, simply match the icon to the one in the bottom corner of the page.

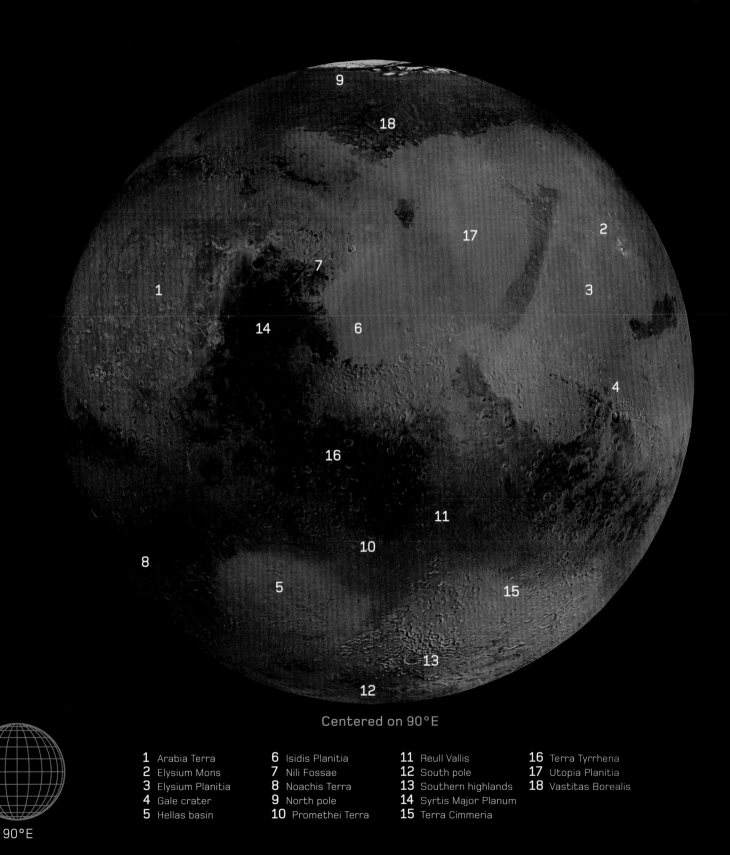

Centered on 90°E

90°E

1 Arabia Terra	**6** Isidis Planitia	**11** Reull Vallis	**16** Terra Tyrrhena
2 Elysium Mons	**7** Nili Fossae	**12** South pole	**17** Utopia Planitia
3 Elysium Planitia	**8** Noachis Terra	**13** Southern highlands	**18** Vastitas Borealis
4 Gale crater	**9** North pole	**14** Syrtis Major Planum	
5 Hellas basin	**10** Promethei Terra	**15** Terra Cimmeria	

65°N

| | MC-02 MC-03
Diacria Arcadia | | MC-04
Mare
Acidalium | MC-05
Ismenius
Lacus | | MC-06
Casius | MC-07
Cebrenia |
30°N

| MC-08
Amazonis | MC-09
Tharsis | MC-10
Lunae Palus | MC-11
Oxia Palus | MC-12
Arabia | MC-13
Syrtis Major | MC-14
Amenthes | MC-15
Elysium |
0°

| MC-16
Memnonia | MC-17
Phoenicus
Lacus | MC-18
Coprates | MC-19
Margaritifer
Sinus | MC-20
Sinus
Sabeus | MC-21
Iapygia | MC-22
Tyrrhenum | MC-23
Aeolis |
30°S

| | MC-24
Phaethontis | MC-25
Thaumasia | MC-26
Argyre | MC-27
Noachis | | MC-28
Hellas | MC-29
Eridania |
65°S

180°W   0°   180°E

← The United State Geological Survey divides Mars into 30 standardized quadrangles, designed to appear rectangular on many map projections. Although they are simply split along set lines of latitude and longitude, each is named after a local geographical feature.

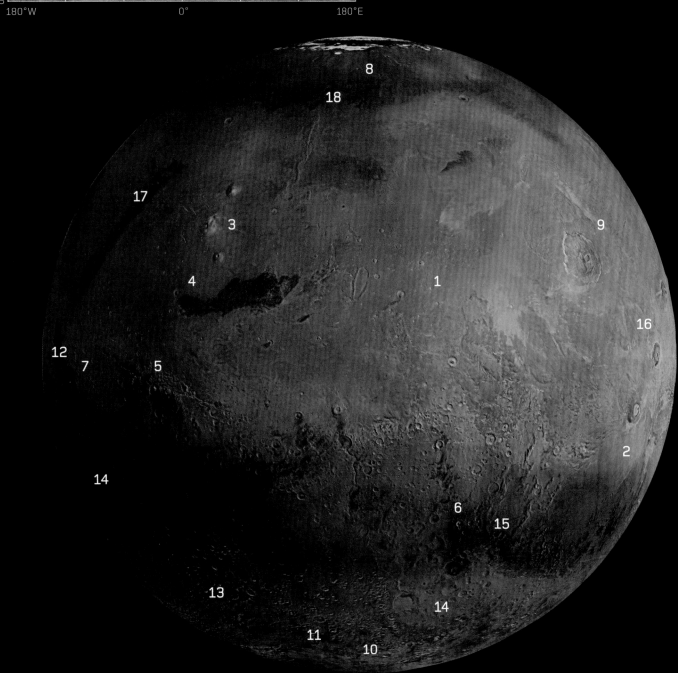

Centered on 180°E

180°E

1 Amazonis Planitia	**6** Gorgonum Chaos	**11** Southern highlands	**16** Tharsis rise
2 Arsia Mons	**7** Isidis Planitia	**12** Syrtis Major Planum	**17** Utopia Planitia
3 Elysium Mons	**8** North pole	**13** Terra Cimmeria	**18** Vastitas Borealis
4 Elysium Planitia	**9** Olympus Mons	**14** Terra Sirenum	
5 Gale crater	**10** South pole	**15** Terra Tyrrhena	

→ The northern and southernmost quadrangles are known as the Mare Boreum and Mare Australe. USGS coordinates frequently measure all locations on Mars in terms of degrees west of the prime meridian, but in this chapter we use degrees east or west for more intuitive interpretation.

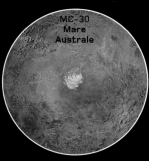

MC-01
Mare
Boreum

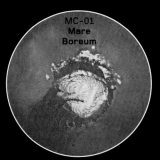

MC-30
Mare
Australe

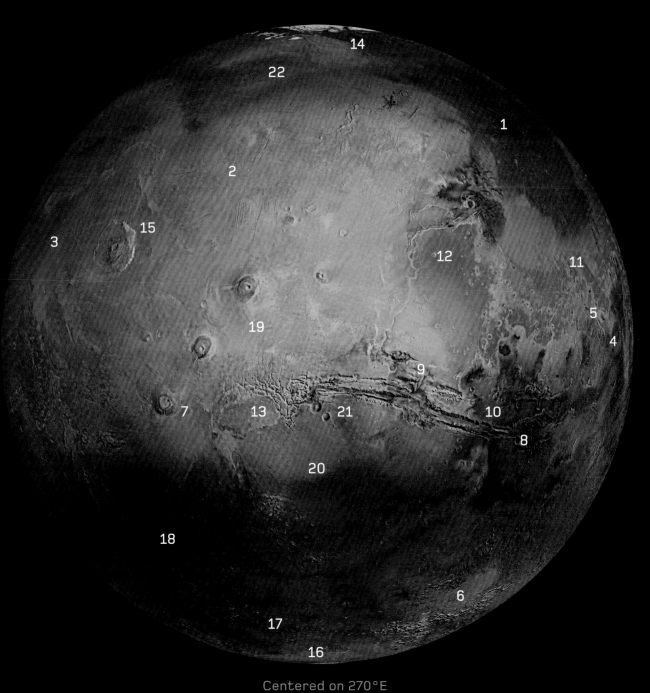

14
22
1
2
15
3
12
11
5
4
19
9
10
7
13
21
8
20
18
6
17
16

Centered on 270°E

270°E

1 Acidalia Planitia	**6** Argyre Planitia	**11** Chryse Planitia	**16** South pole	**21** Valles Marineris
2 Alba Mons	**7** Arsia Mons	**12** Kasei Vallis	**17** Southern highlands	**22** Vastitas Borealis
3 Amazonis Planitia	**8** Aureum Chaos	**13** Noctis Labyrinthus	**18** Terra Sirenum	
4 Aram Chaos	**9** Candor Chasma	**14** North pole	**19** Tharsis rise	
5 Ares Vallis	**10** Capri Chasma	**15** Olympus Mons	**20** Thaumasia	

Valles Marineris
13.9°S, 59.2°W

The single most prominent surface feature on Mars is undoubtedly the enormous canyon system that runs around almost an entire hemisphere, south of the Martian equator. Named in honor of the Mariner 9 Mars orbiter, which sent back the first detailed photographs of it in the early 1970s, the Valles Marineris, or Mariner Valleys, extend for some 2,500 miles (4,000 kilometers)—the width of the continental United States. Up to 370 miles (600 kilometers) wide at their broadest and some 6 miles (10 kilometers) deep in places, they dwarf Earth's own Grand Canyon and are one of the largest landscape features in the entire solar system.

Despite their name, however, the Valles Marineris do not owe their origin to water—no river ever carved this valley, though water may later have gathered on its floor and even eroded its sides. Fundamentally, this is a geological fault—a huge crack in the Martian surface opened up by shifts in the crust on either side.

The exact nature of these shifts is still open to debate: for a long time, astronomers assumed that the canyon formed in response to the growth of the Tharsis rise that lies directly to its north. As huge amounts of volcanic rock piled up in this region, the nearby crust buckled under the strain and split apart to open a vast fissure. In contrast, recent studies of the Valles have suggested an origin that seems more familiar from Earth: in their broadest region, known as Candor Chasma, there may be evidence to suggest the two sides of the valley have slipped horizontally, moving roughly 93 miles (150 kilometers) relative to each other in a roughly east–west direction. This is currently the strongest evidence supporting the idea that Mars may be tectonically active after all (see page 28).

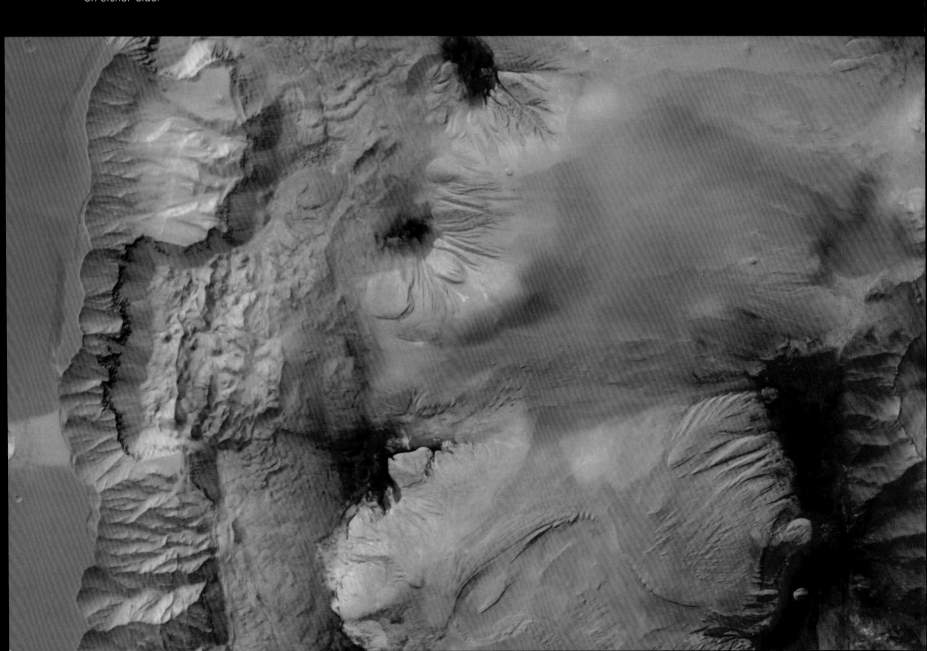

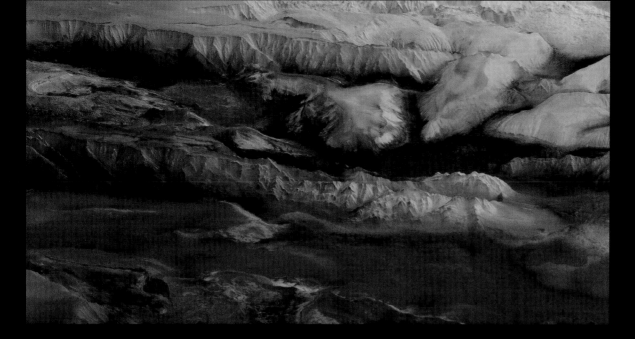

← This perspective view of the central Valles Marineris is compiled using data from the European Mars Express probe. In this area, the canyon splits into three separate valleys—Melas Chasma, Candor Chasma (see page 142), and Ophir Chasma.

↓ A Mars Express view focuses on details within Ophir Chasma. Cliffs at far left plunge up to 3.1 miles (5 kilometers) down to the canyon floor, and show clear signs of sculpting by heavy erosion and massive landslips. Huge piles of eroded material form hills within the valley, while there are also signs that the region has been shaped by both volcanic activity, glaciers, and flowing water.

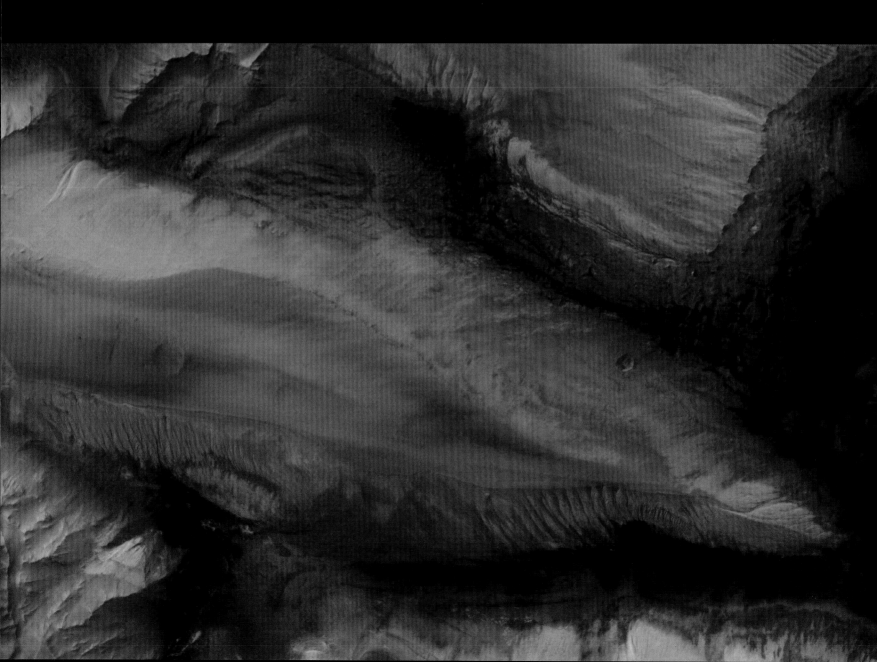

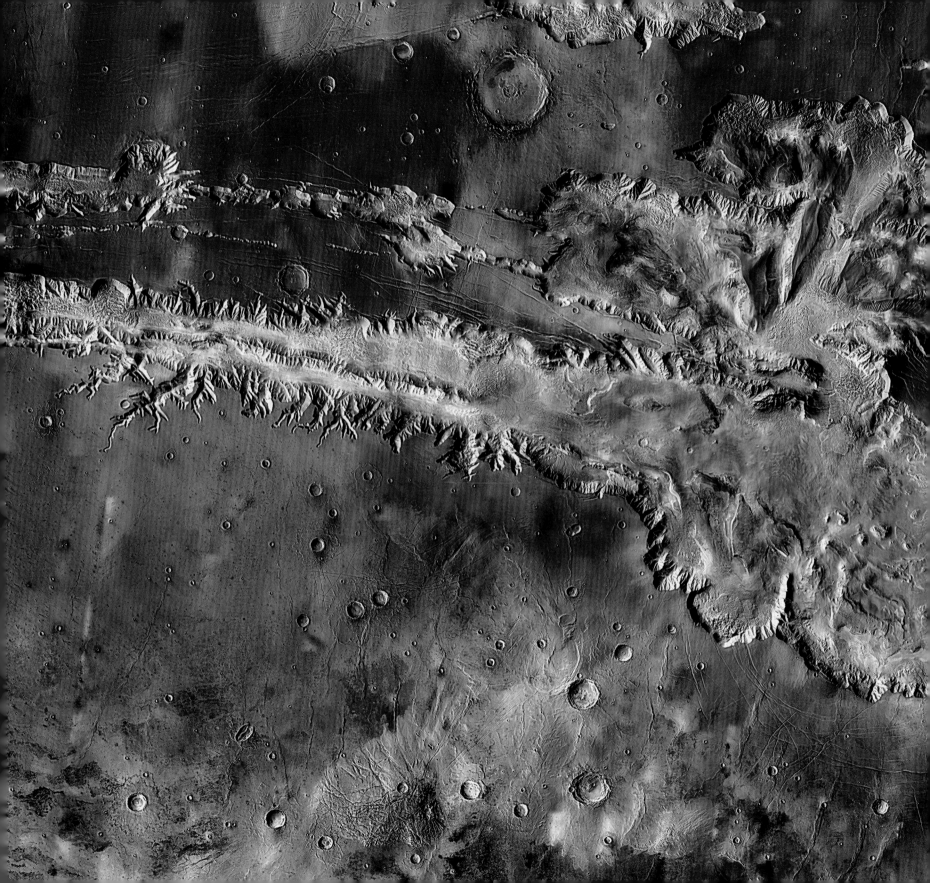

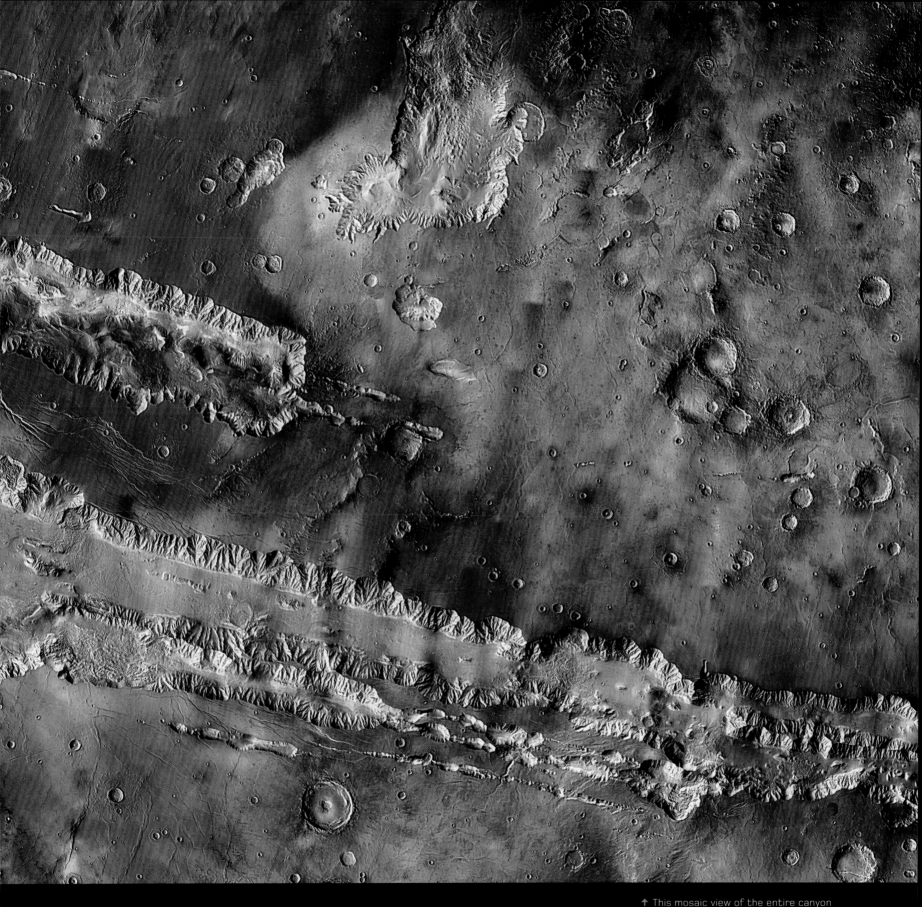

↑ This mosaic view of the entire canyon complex is composed of more than 500 infrared images from the Thermal Emission Imaging System (THEMIS) aboard NASA's Mars Odyssey spacecraft. Despite the enormous extent of the image, it captures details as small as a soccer field.

Aram Chaos

2.6°N, 21.5°W

This large feature just north of the Martian equator is barely recognizable as the remains of an ancient impact crater that have suffered heavy erosion and transformation since. The crater originally had a diameter of around 174 miles (280 kilometers), but it is now little more than a shallow depression in the landscape. As its name suggests, the original crater floor is now a chaotic jumble of flat-topped blocks and rounded hummocks separated by broad valleys.

The features of Aram Chaos suggest dramatic erosion at work—not from a passing torrential flood as seen, for example, in the Kasei Valles (see page 136), but from within. The floor of the crater seems to have subsided and collapsed, most likely as layers of ice or water were removed from beneath the surface. Researchers believe that the water became trapped in the crater along with layers of wind-blown debris that created sedimentary rock over millions of years after the crater formation.

As the Martian climate altered, it was transformed into an icy permafrost, but then catastrophically melted, welling to the surface to create a deep lake as the ground around it collapsed. This temporary lake eventually drained away through Aram Vallis, a narrow channel on the crater's eastern side. Another phase of sedimentation followed, perhaps around 2.7 billion years ago, and the presence of the mineral hematite in the crater suggests that this also took place in a wet environment.

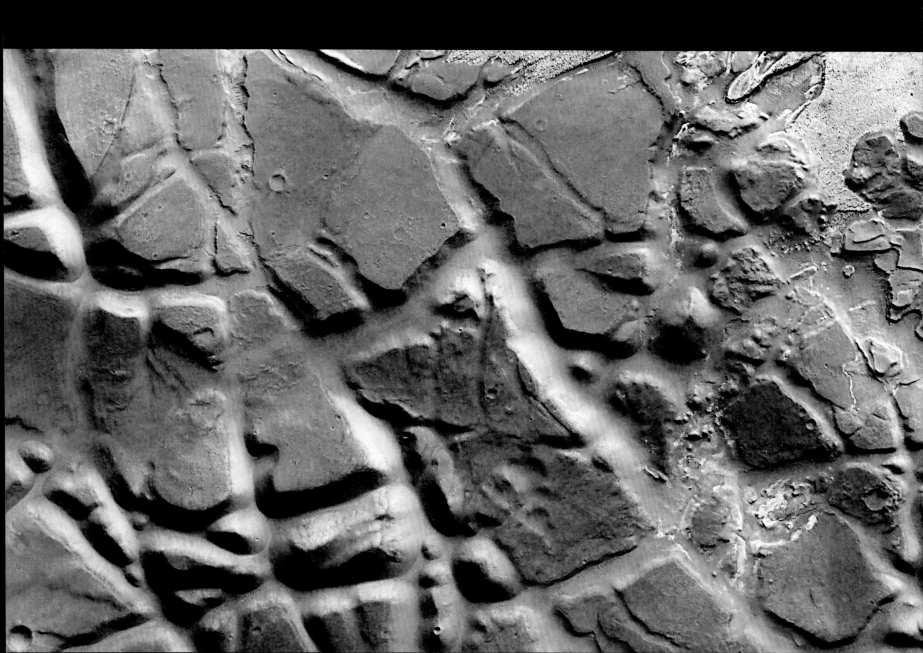

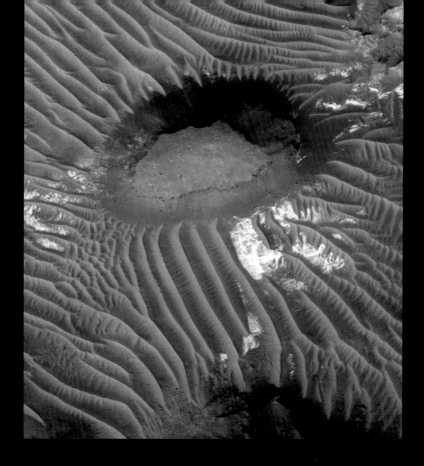

↓ This false-color image from the High-Resolution Stereo Camera aboard Mars Express reveals the scale and complexity of the Aram Chaos landscape—a confusing jumble of subsided blocks, eroded mounds, flood-carved channels, and (near the top of the image) bright, smooth areas that seem to be later deposits of sedimentary rock.

← An intricate dune field surrounds a mesalike outcrop in the eastern part of Aram Chaos, close to its outflow into nearby Ares Vallis (see page 156). The smooth, dark apron surrounding the mesa is probably dark material eroded directly from its slopes, while the dust that makes up the surrounding dunes may well have been transported from elsewhere.

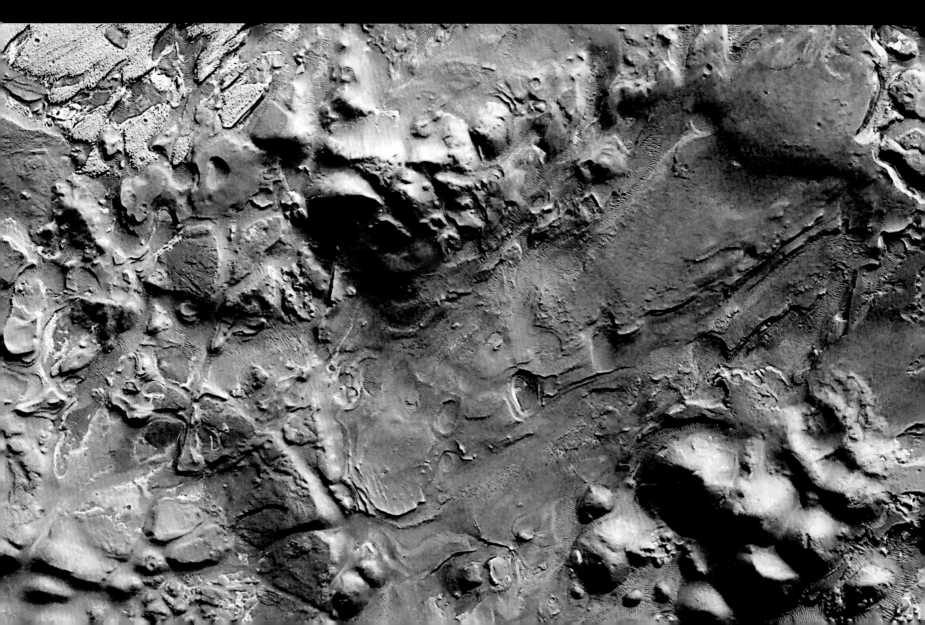

Tharsis rise
1.6°N, 112.6°W

Dominating the western hemisphere of Mars, Tharsis is an enormous volcanic plateau, rising some 23,000 feet (7,000 meters) above the average Martian datum. It plays host to some of the largest volcanoes in the solar system.

Named after the biblical land of the uttermost west, the region stretches across roughly 4,350 miles (7,000 kilometers) from north to south, and 3,100 miles (5,000 kilometers) from east to west. Its boundaries are poorly defined, but it is usually taken to consist of a large oval area (the southern rise) extending from the Noctis Labyrinthus and Valles Marineris in the east, and underlying a large area of the southern highlands. A lobe to the northwest (the northern rise) extends into the planet's northern plains and encompasses another huge volcano, Alba Mons (see page 116).

Olympus Mons, officially the biggest single volcano in the solar system, is situated just off the plateau's western slope.

At the heart of the southern rise lies a chain of three very similar volcanoes running roughly from northeast to southwest—Ascraeus, Pavonis, and Arsia Montes. Each is a large volcanic shield in its own right, and in terms of volume, Arsia is in fact the second biggest volcano on Mars after Olympus Mons. The distribution of these volcanoes recalls volcanic chains found on Earth, such as the Hawaiian Islands, and suggests that they may be formed in a similar way: a plume of hot material rising through the mantle below creates a magma-producing hot spot at the surface that drifts over time with the movement of the overlying crust.

↓ A topographic map of the Tharsis region, based on data from the Mars Orbiter Laser Altimeter (MOLA) instrument aboard NASA's Mars Global Surveyor spacecraft, reveals the huge scale of the Red Planet's largest volcanic province. Some scientists have argued that the entire region should be seen as a single supervolcano with the central chain of the Tharsis Montes at its peak.

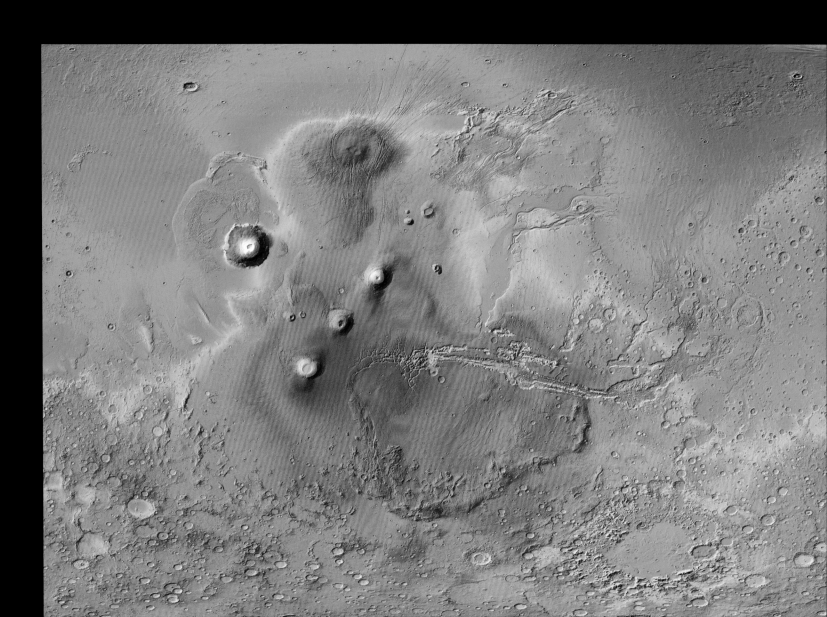

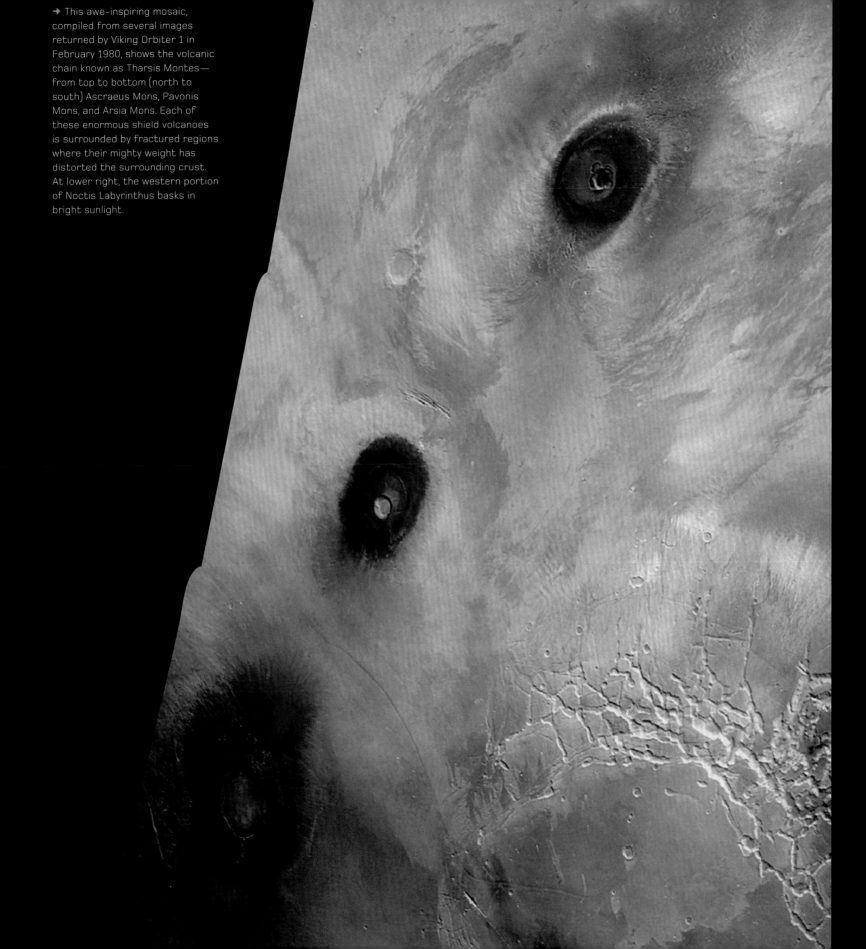

➜ This awe-inspiring mosaic, compiled from several images returned by Viking Orbiter 1 in February 1980, shows the volcanic chain known as Tharsis Montes— from top to bottom (north to south) Ascraeus Mons, Pavonis Mons, and Arsia Mons. Each of these enormous shield volcanoes is surrounded by fractured regions where their mighty weight has distorted the surrounding crust. At lower right, the western portion of Noctis Labyrinthus basks in bright sunlight.

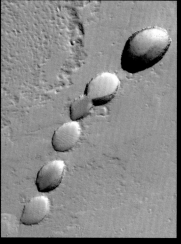

← This chain of pits on the lava plains to the north of Ascraeus Mons, known as a catena, is a result of collapse from below rather than a meteorite impact. Such collapse pits are thought to form when tectonic forces pull at opposite sides of a volcanic plain, or when the roof of an empty subterranean tunnel or lava tube caves in.

↓ Tadpole-like valleys, running across the flanks of Pavonis Mons, are formed by lava tubes whose roofs have entirely collapsed. Such tubes are created when the surface of a volcanic plain solidifies, but lava continues to run beneath.

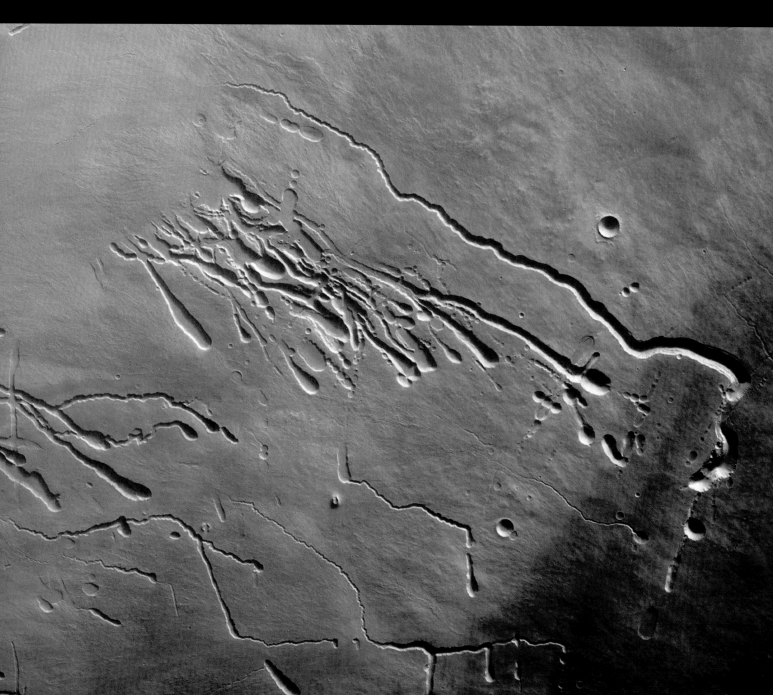

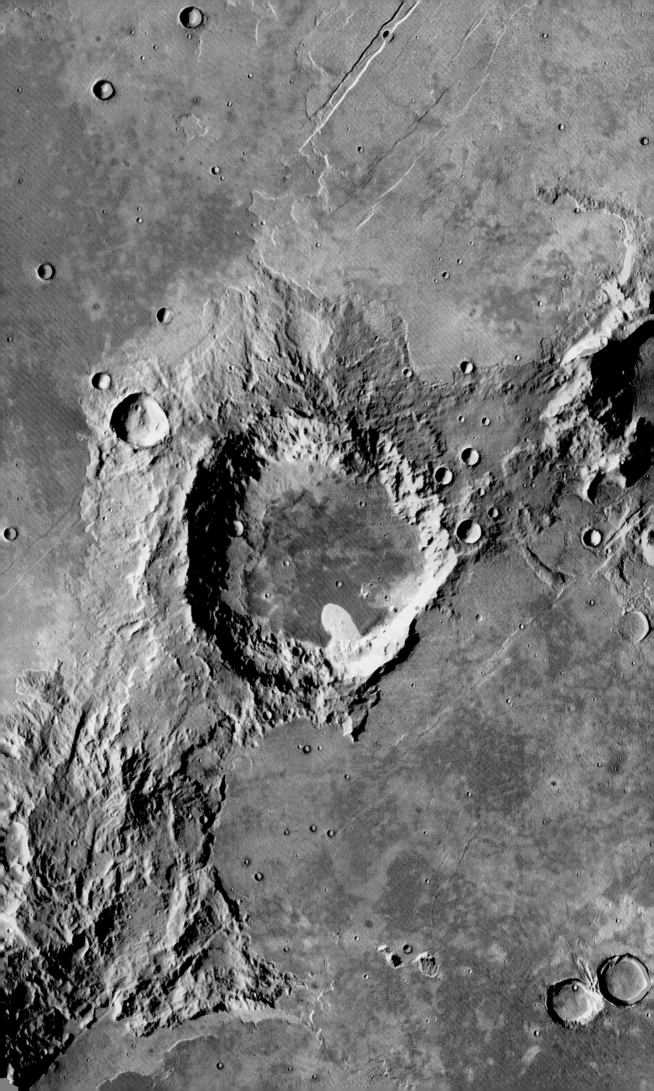

Olympus Mons
18.7°N, 133.8°W

The tallest volcano in the solar system, Olympus Mons lies just to the west of the Tharsis rise, and soars to an awe-inspiring 13.2 miles (21.2 kilometers) above the average Martian surface datum. With a diameter of 388 miles (624 kilometers), it is roughly the same size as the US state of Arizona and has 100 times the volume of Mauna Loa, Earth's largest volcano. Much of its outer edge is rimmed by cliffs that plunge up to 3.7 miles (6 kilometers) to the surrounding terrain, while its peak is marked by a huge caldera complex—a series of pitlike depressions miles deep, formed as the magma chamber deep beneath the visible volcano finally drained away.

Olympus Mons, like all its giant Martian siblings, is a shield volcano—an enormous mountain of volcanic lava created by repeated eruptions along its flanks. Traditionally, the huge size of Martian volcanoes has been attributed to the planet's lack of tectonics—on Earth, the crust is constantly (if gradually) moving and rearranging itself, so the geological factors that can trigger volcanic activity rarely stay in the same place for long. The hot spots that trigger Martian volcanoes, in contrast, should remain effectively fixed in place for many millions of years, allowing enormous volcanoes to build up through repeated eruptions. However, the apparent movement of a volcanic hot spot beneath the Tharsis rise casts doubt on the idea that the Martian crust is entirely static, and the recent discovery of tectonic faults (see page 28) adds to the complications. As a result, it seems likely that Olympus Mons owes its mighty scale to a combination of different factors.

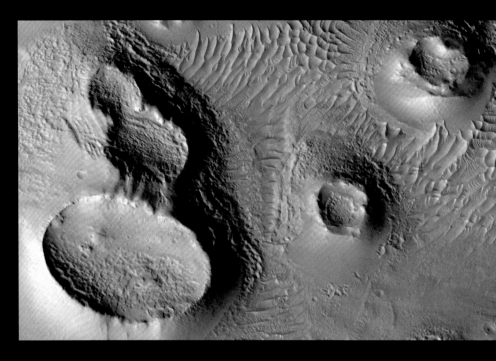

↑ This field of relatively small volcanic cones lies to the north of the main shield of Olympus Mons. As this HiRISE image shows, the entire region is blanketed in thick layers of dust, as is Olympus Mons itself.
↓ The graphic below shows the enormous scale of Olympus Mons compared to Earth's highest mountain, Everest.

→ This spectacular overhead view of Olympus Mons is a mosaic compiled from Viking 1 orbiter images. It shows the huge extent of the volcanic shield, and the surrounding lava plains, covered in complex ridged patterns called sulci that are thought to have formed as a result of massive landslides on the volcano's flanks.

Olympus Mons
Height: 69,649 feet
(21,229 m) above datum
Diameter: c.372 miles
(600 kilometers)

Mount Everest
Height: 29,029 feet
(8,848 m) above sea level
Diameter: c. 125 miles
(200 kilometers)

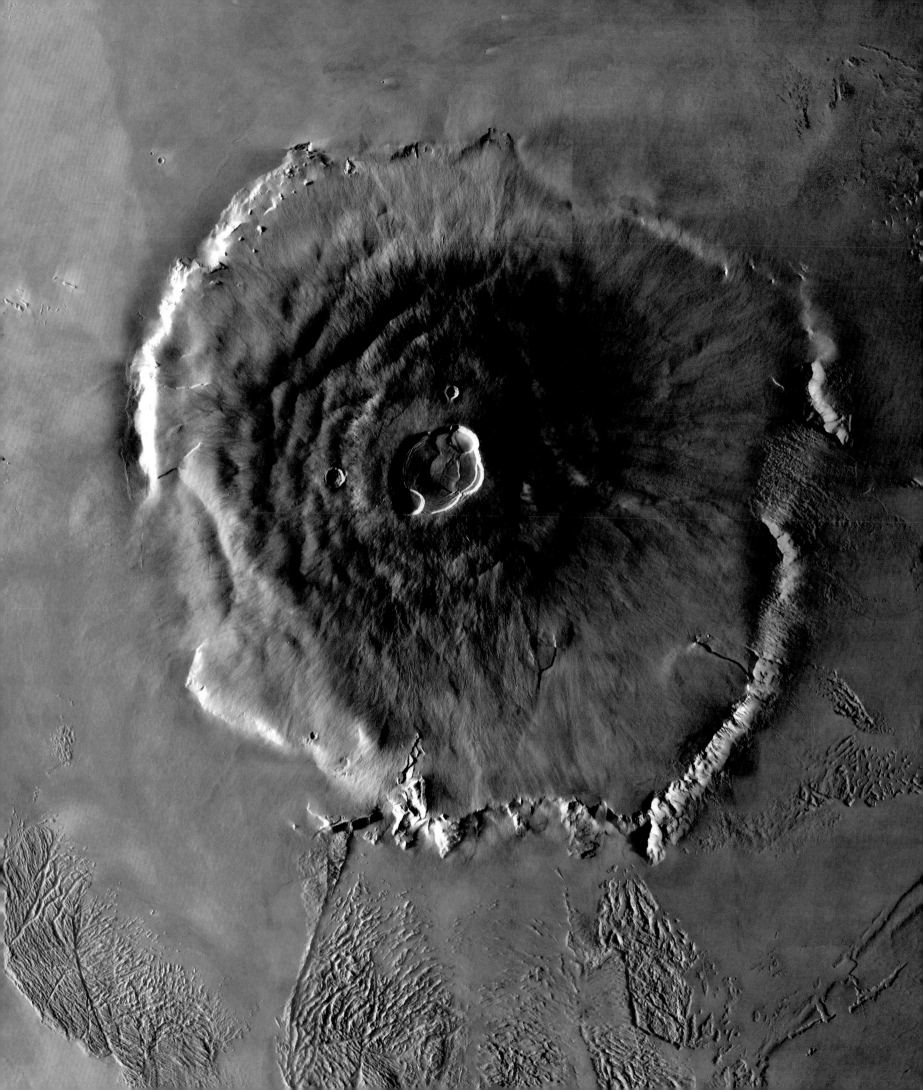

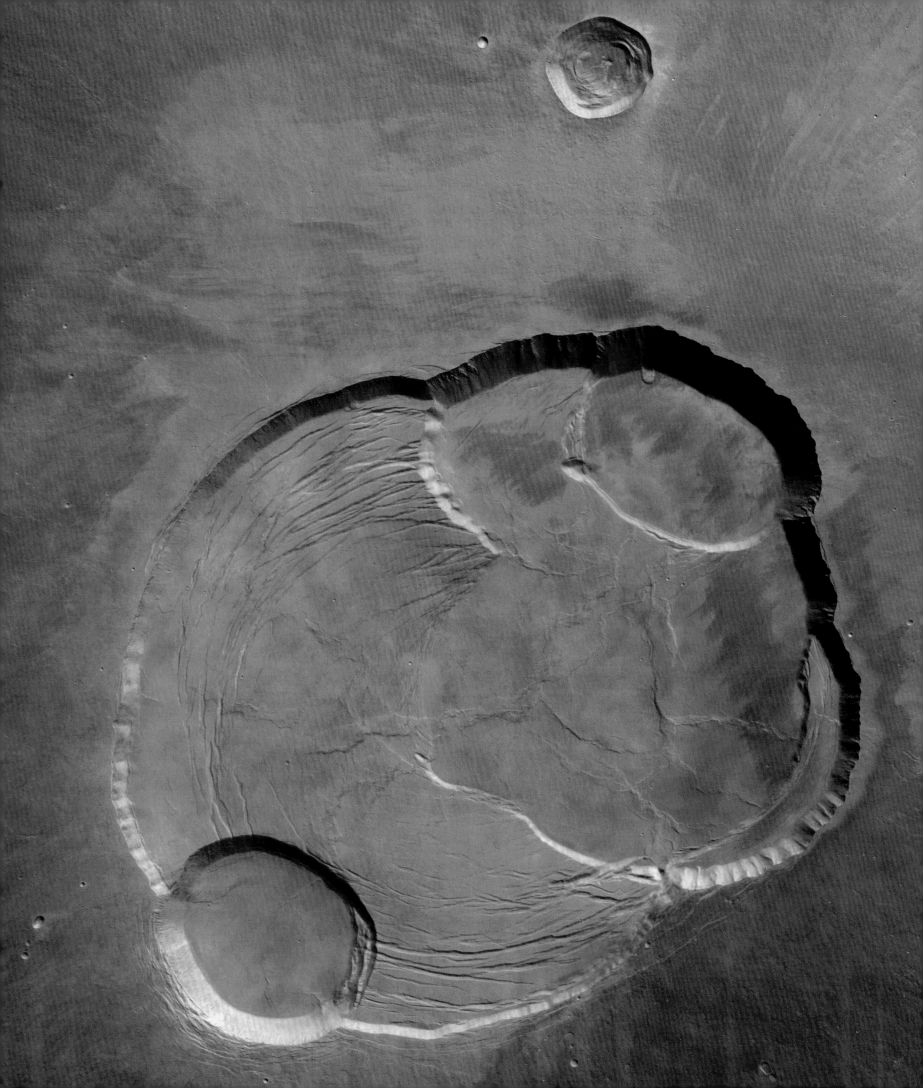

← This Mars Express view shows complex detail in Olympus's central caldera. Here, the summit of the volcanic shield has subsided in several stages, leaving cliffs up to 1.9 miles (3 kilometers) tall around a series of overlapping depressions. The entire image is roughly 62 miles (100 kilometers) wide.

→ A topographic map of Olympus Mons from the Mars Global Surveyor uses colors to indicate different elevations from the blue and green of the underlying plateau to the red and white of the highest regions around the caldera. A steep escarpment surrounds the volcano on its northwest and southeast sides, while in other areas the lava merges more smoothly into the surrounding plain.

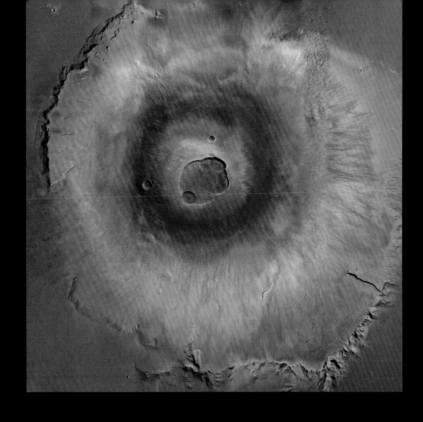

↓ A Mars Express image shows complex features around the volcano's eastern scarp, where cliffs tower up to 3.7 miles (6 kilometers) high. Fracture-like patterns were probably torn open by tectonic stresses as the crust deformed under the volcano's enormous weight, while in other places crisscrossing channels suggest the action of water, perhaps escaping from just beneath the surface. Based on crater counts on the various terrains, the boundary zone seems to have been active up to 30 million years ago—comparatively recently in Martian terms.

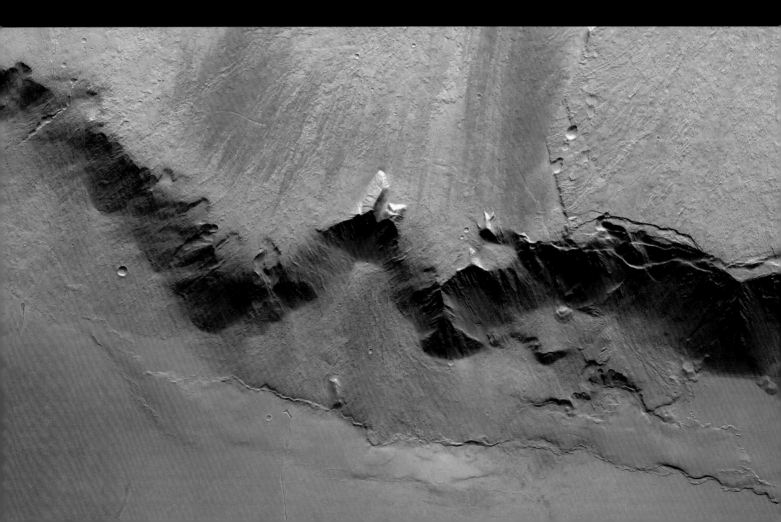

North pole
90°N

While much of the Martian northern hemisphere is dominated by a vast lowland plain, the north polar ice cap lies on a high plateau known as the Planum Boreum, 750 miles (1,200 kilometers) across. This enormous dome of compressed earth and ice rises up to 1.9 miles (3 kilometers) above the surrounding Vastitas Borealis, and displays a complex structure of spiral troughs that often cut deep into the surface ice.

The permanent polar cap consists largely of water ice, with a seasonal frost of frozen carbon dioxide that extends across the Vastitas Borealis and vastly increases its overall extent and visibility. The spiral features are thought to be a result of a complex process in which winds blowing away from the polar highlands develop a spiral pattern thanks to the planet's rotation (the Coriolis effect), and transport sublimated ice away from the sunlit slopes. The troughs are thought to actually run perpendicular to the direction of the winds themselves.

The bulk of the Planum Boreum consists of complex sedimentary rocks known as polar layered deposits. These are thought to have built up out of atmospheric dust trapped as new layers of carbon dioxide frost are deposited each fall. Although much of the gas sublimates back into the atmosphere the following spring, some may be left behind along with the dust. In theory, therefore, these layers should contain a trapped record of changes to the Martian atmosphere over many millions of years.

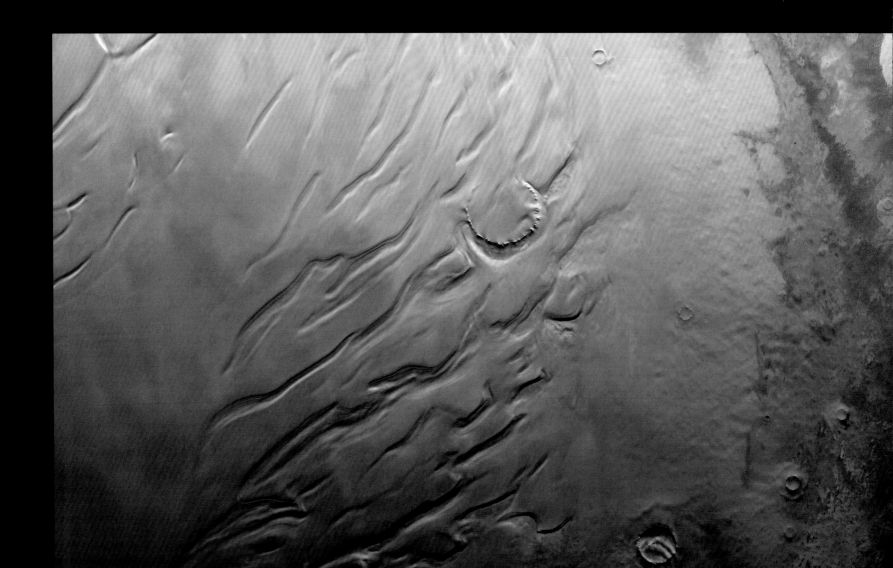

↓ This spectacular image of the edge of the north polar ice cap was captured by the High-Resolution Stereo Camera (HRSC) aboard ESA's Mars Express spacecraft. It shows the boundary zone between the dune-filled plains of the Vastitas Borealis on the right (see also page 110) and the raised polar plateau of the Planum Boreum to the left. The plateau shows a clear pattern of spiral canyons etched into its layered ice by so-called "katabatic" winds (see page 40).

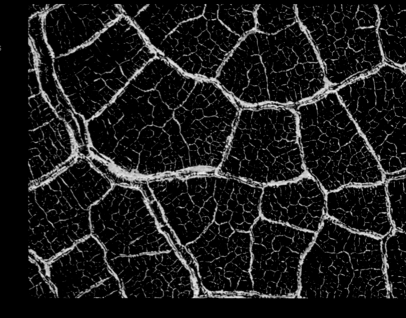

↑ Resembling an enormous honeycomb pattern, this complex web of bright filaments at the edge of the polar cap shows areas where repeated seasonal expansion and contraction of the ground has produced either ridges or troughs that accumulate bright frosts more easily than their surroundings.

↓ This perspective view of the polar ice cap, generated from Mars Express observations, reveals towering, scallop-edged cliffs up to 1.2 miles (2 kilometers) high at the edge of the ice-covered plateau. Dark material accumulating at the foot of the cliffs and in the foreground dune field is likely to be volcanic ash.

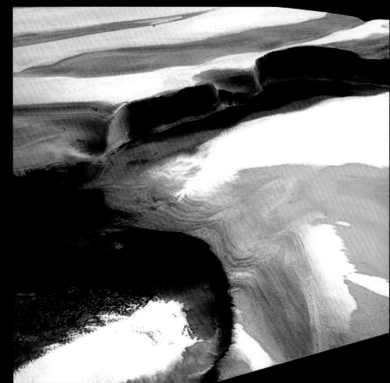

Hellas basin
43°S, 70°E

Situated amid the southern highlands, the Hellas basin is one of the most prominent features on Mars, thanks to the presence within it of a bright, flat plain known as Hellas Planitia. Thanks to its size and brightness, it was one of the first real features to be identified on Mars, as early as 1867. It owes its current name, from the Greek word for Greece itself, to the great Italian observer of Mars, Giovanni Schiaparelli (see page 14).

Hellas is a giant impact crater—the largest so far confirmed to exist on Mars, with a roughly circular shape and a diameter of about 1,430 miles (2,300 kilometers). Between the mountainous rim and the basin floor, there is an elevation difference of some 5.6 miles (9 kilometers), with the floor some 23,000 feet (7,000 meters) below the Martian datum. This makes Hellas Planitia the lowest part of Mars, and consequently gives it the highest atmospheric pressure—high enough in places to allow liquid water to persist on the surface. Although there is no actual evidence for the presence of liquid water at present, Hellas's location amid the highlands has made it a natural drainage basin throughout its history, and there is strong evidence for past water flowing onto the plain and even shaping the crater floor. Glacierlike structures have also been found within Hellas, suggesting that frozen water still exists and flows just beneath the surface layers of dust.

→ Features such as this so-called "tongue glacier" are found in many places across Hellas Planitia, and although they are covered in the ubiquitous Martian dust, they are almost certainly formed by slow-moving ice flows. Elsewhere, ice has been directly detected in "lobate debris aprons" within several other small craters within Hellas.
↓ This view from the HRSC camera aboard Europe's Mars Express probe shows a strip of the "transition region" on the northern edge of the basin. Here, a highland region called Tyrrhena Terra merges with the bright surface of Hellas Planitia visible at extreme right. Rich in hydrated minerals, this region was shaped by the presence of water for a large part of its history and probably drained into Hellas.

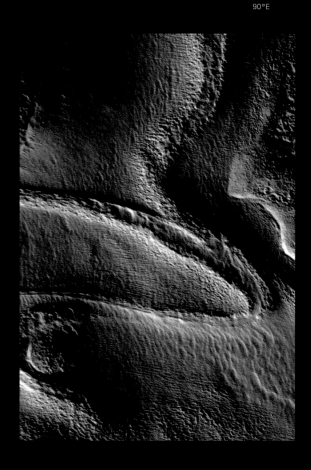

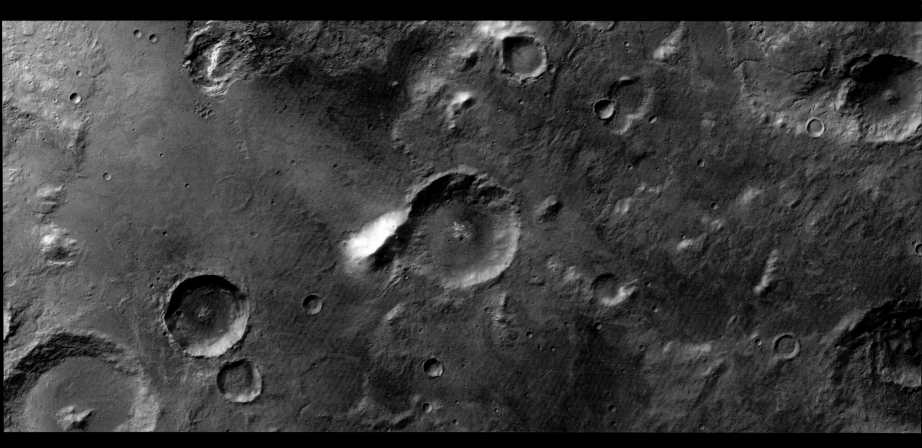

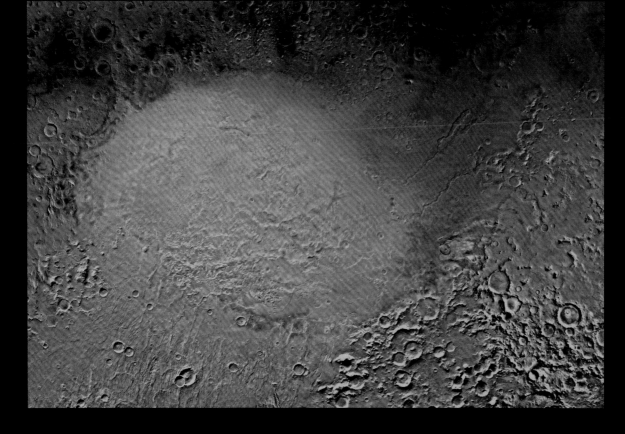

← The Hellas basin forms a bright, smooth oval surrounded by darker, rougher terrain in this large-scale mosaic compiled from Viking orbiter images. While the edges of this enormous crater are clearly defined along its northwestern and southeastern rims, in other parts the rim is eroded and buried by later deposits (most likely from nearby volcanoes). Nevertheless, the presence of channels that once carried water from the surrounding highlands can clearly be traced in several places.

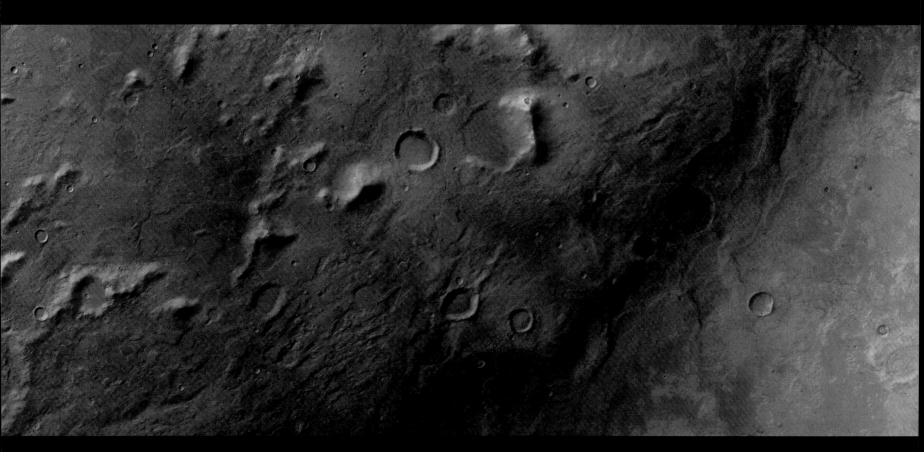

Southern highlands
c.0°–70°S

Heavily cratered highland terrain covers roughly two-thirds of the surface of Mars, including the entire southern hemisphere, creating a stark dichotomy with the northern lowland plains. The Martian highlands are a complex landscape that still bears the scars of a history that began before the Late Heavy Bombardment—the period around 3.9 billion years ago when comets and asteroids scattered by the motion of the gas giant planets bombarded the solar system's vulnerable inner worlds. Although the highlands have been significantly modified by events in later Martian history, they have never been recycled by the processes of plate tectonics in the way that almost all of Earth's crust has been.

In part, this may be due to the thickness of the underlying planetary crust—at up to 36 miles (58 kilometers) deep it is almost twice as deep as the crust beneath the northern plains, and the buoyancy of its deep roots may help to push the highlands up to an average of 1.9–3.8 miles (3-6 kilometers) higher than the lowland regions. Amid the countless craters, the highlands show signs of many other Martian processes at work, including the paths of ancient rivers and sudden dramatic floods, chaotic terrain, and present-day glaciation.

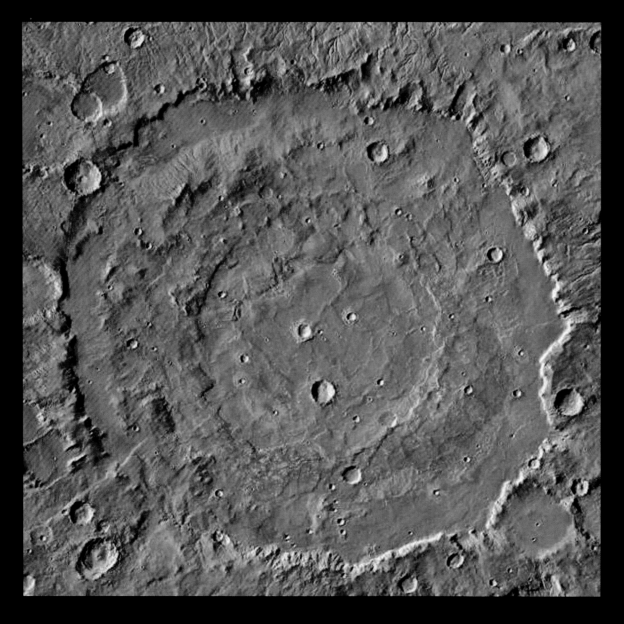

← Huygens is a 280-mile (450-kilometer) basin in the southern highlands, whose formation several billion years ago exposed minerals buried up to 3 miles (5 kilometers) deep in the Martian crust. While much of the rock has subsequently been re-covered by later activity such as the accumulation of dust, these ancient layers can be exposed for a second time by the formation of later craters within the basin. This is precisely the process that has taken place in the small crater at roughly "ten o'clock" on the rim of Huygens, where buried carbonate minerals have been reexposed (see center right picture opposite).

➜ Promethei Terra is a large highland region to the east of Hellas Planitia. Here, large amounts of dust have accumulated over billions of years, softening and smoothing its rugged features at all but the sharpest edges. The large impact crater to the right of center is 20 miles (32 kilometers) wide, and has a dark center where the crater floor has been uplifted and eroded.

◥ This HiRISE image shows part of Huygens crater (see opposite) just 1,500 feet (460 meters) across. It reveals an intricate network of fractures, where clay and carbonate minerals have been detected. The carbonates probably formed through the long-term action of water, and would have absorbed significant amounts of carbon dioxide, contributing to the thinning of the Martian air.

➜ This colorful view from Europe's Mars Express mission shows elevation differences in Ulyxis Rupes, a boundary region between smooth, dust-covered polar ice and the more rugged highlands (toward the right). Glacierlike features are prominent at the center of the image. The same area is shown in stunning true-color overleaf.

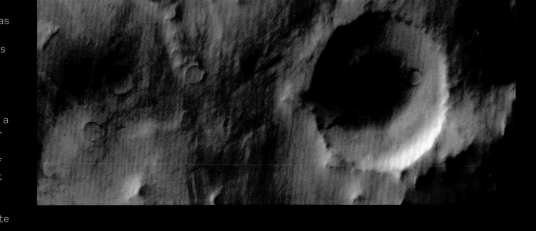

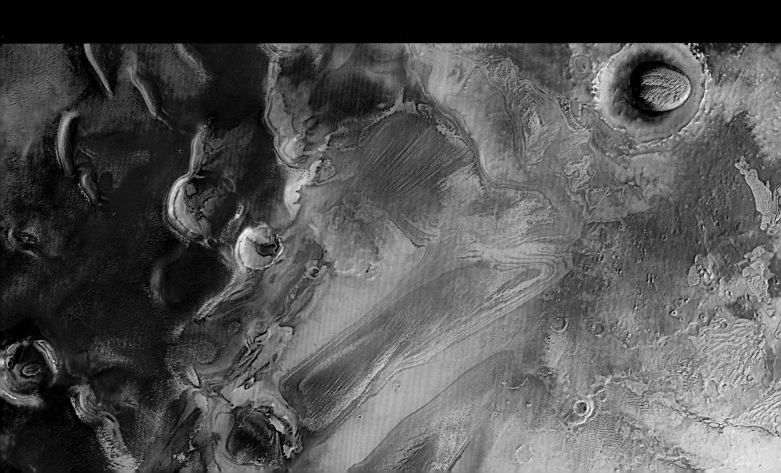

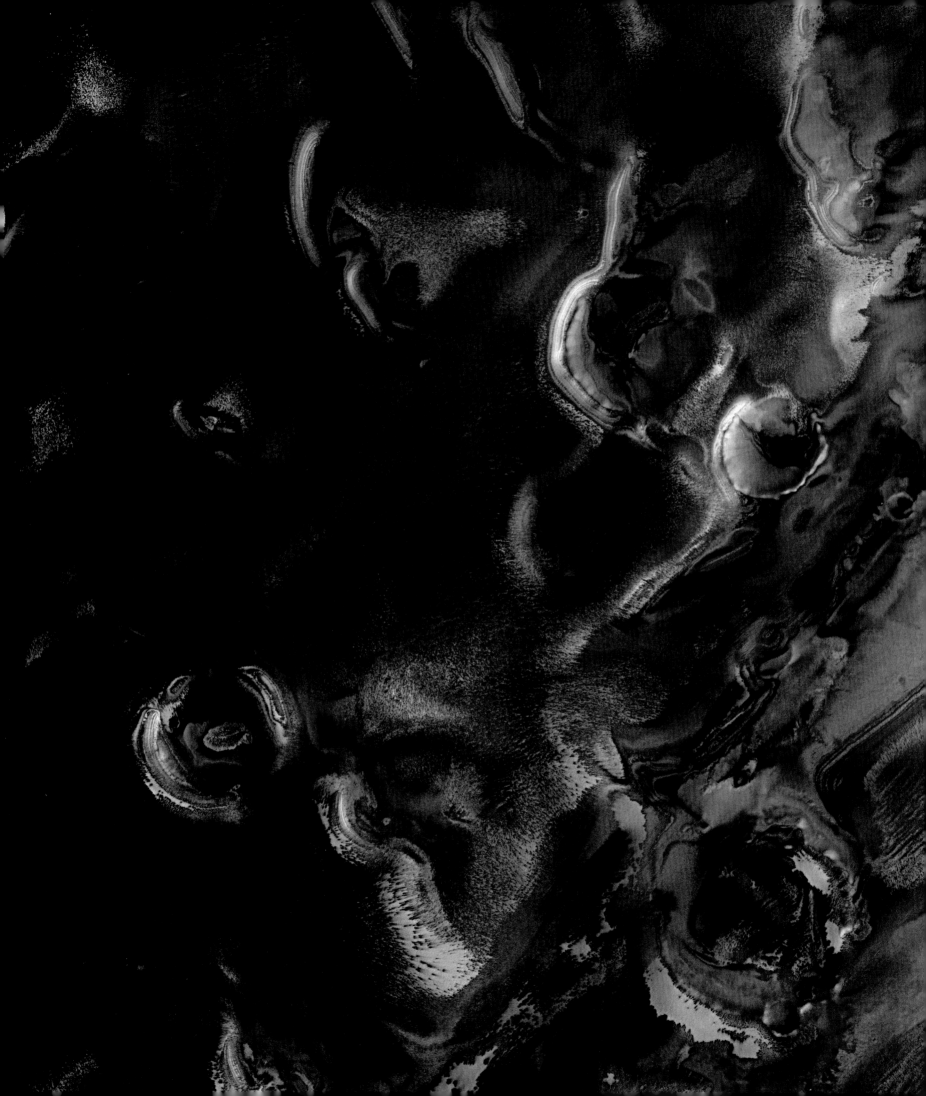

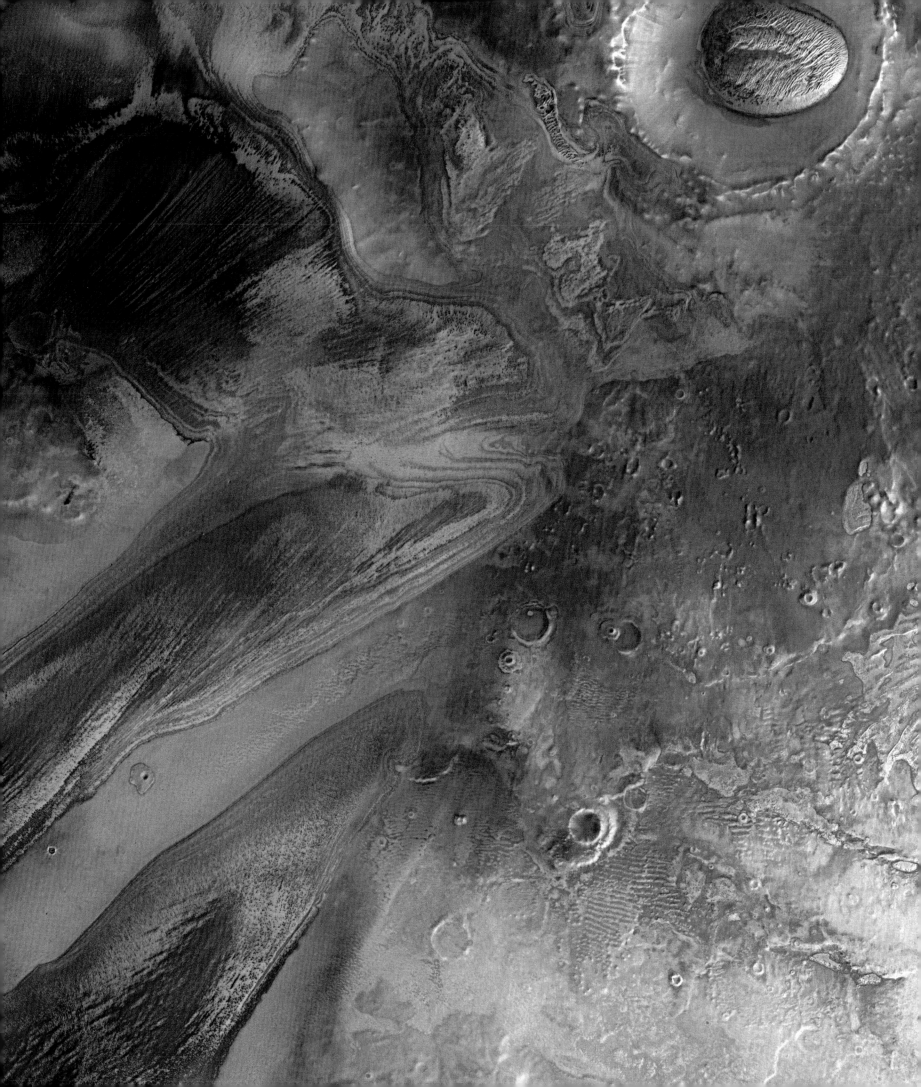

0°

The south polar ice cap, rather like its northern counterpart, lies on top of an elevated plateau. The visible permanent cap is considerably smaller than the northern ice cap, with a diameter of just 250 miles (400 kilometers), and is centered not on the pole, but on a point 90 miles (150 kilometers) into the western hemisphere. This strange situation is believed to be a result of local weather systems created by the Hellas basin triggering increased snowfall on the western side of the pole. However, the extensive seasonal ice cap, whose carbon dioxide frosts can reach a thickness of up to 40 inches (1 meter), is centered on the geographic south pole itself. In 2007, ground-penetrating radar confirmed the long-suspected presence of huge amounts of water ice buried just beneath the surface of the south polar cap—enough, it seems,

to cover the entire planet to a depth of more than 33 feet (10 meters).

Although the eroded patterns in the southern ice cap are superficially similar to those in the northern hemisphere, they show less of a clear spiral structure, most likely due to the more complex weather conditions generated by the surrounding southern highlands. Evidence of Martian climate change (see page 66) is also most obvious among the pitted surfaces of the south polar cap, and it seems likely that the cap's extent and thickness varies dramatically with the Martian climate cycle.

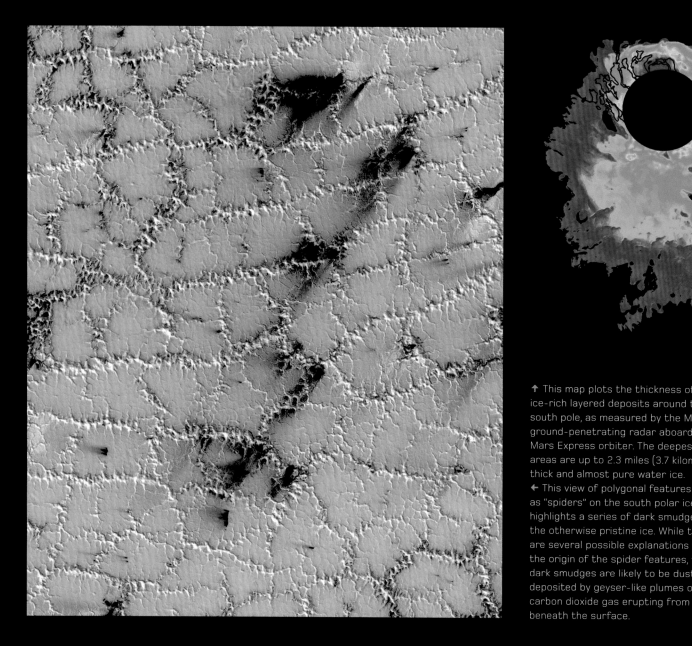

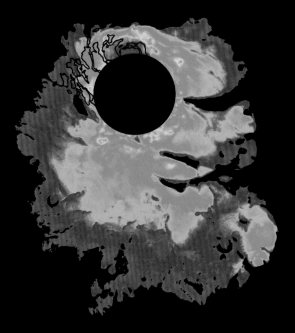

↑ This map plots the thickness of ice-rich layered deposits around the south pole, as measured by the MARSIS ground-penetrating radar aboard the Mars Express orbiter. The deepest red areas are up to 2.3 miles (3.7 kilometers) thick and almost pure water ice.
← This view of polygonal features known as "spiders" on the south polar ice cap highlights a series of dark smudges on the otherwise pristine ice. While there are several possible explanations for the origin of the spider features, the dark smudges are likely to be dust deposited by geyser-like plumes of carbon dioxide gas erupting from beneath the surface.

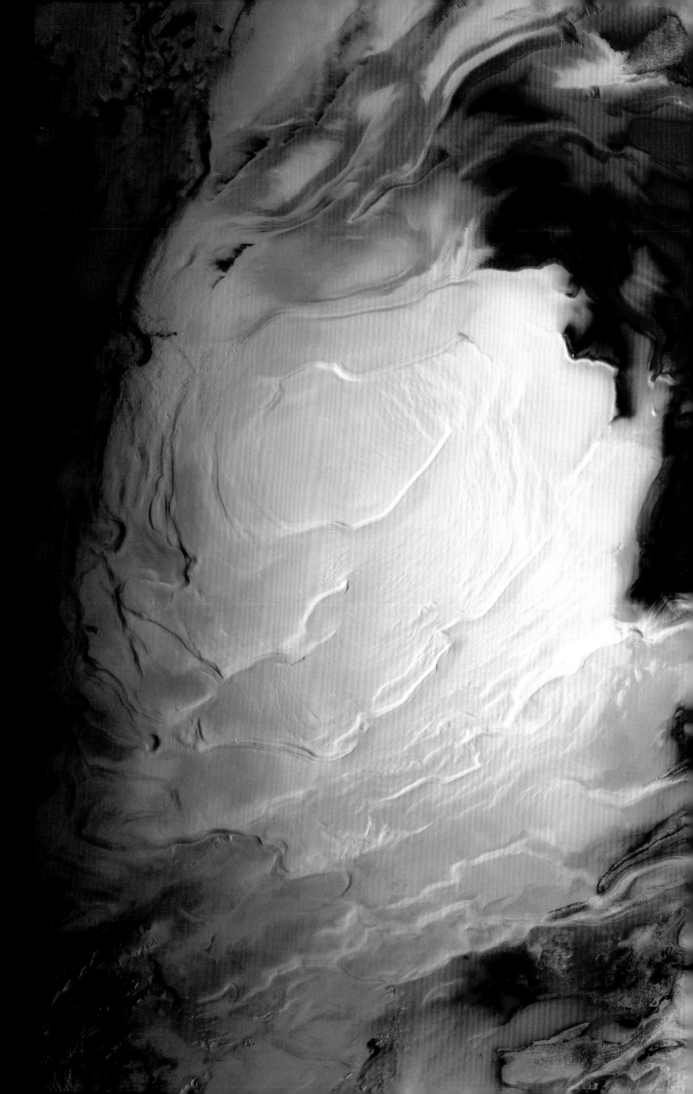

→ This beautiful view of the region around the south polar cap was put together from infrared, green, and blue images gathered by the European Space Agency's Mars Express orbiter. Exposed brilliant white ice in the middle marks the visible polar cap, but this image highlights the gradual buildup of dust on the polar plateau's outer flanks, gradually burying the ice beneath deeper and deeper layers.

A 104-mile (167-kilometer) crater in the Arabia Terra region of the northern hemisphere, Becquerel is named after the French physicist and codiscoverer of radioactivity Henri Becquerel. In 2001, it was the site of a spectacular discovery in the form of complex multilayered deposits within the crater bedrock. There are many dozens of layers in total, ranging in thickness between an average of 11.8 feet (3.6 meters) in some places and 118 feet (3.6 meters) in others. They clearly formed through sedimentation—the deposition and gradual compression of small particles—and the cyclical nature of their formation strongly suggests that they were mostly formed in a standing body of water that accumulated and either drained away or evaporated on several occasions. In many places, the layers accumulated in groups of ten—a pattern that researchers have linked to periodic variations in the Martian climate (see page 66).

Another significant feature, revealed in wide-angle color images, is Becquerel's color variation. Dark streaks were probably produced by volcanic dust blown across the landscape and dropped where it was protected from prevailing winds by the crater walls, but an intriguing light-colored mound is thought to be an outcrop of sedimentary rock rich in sulfate minerals.

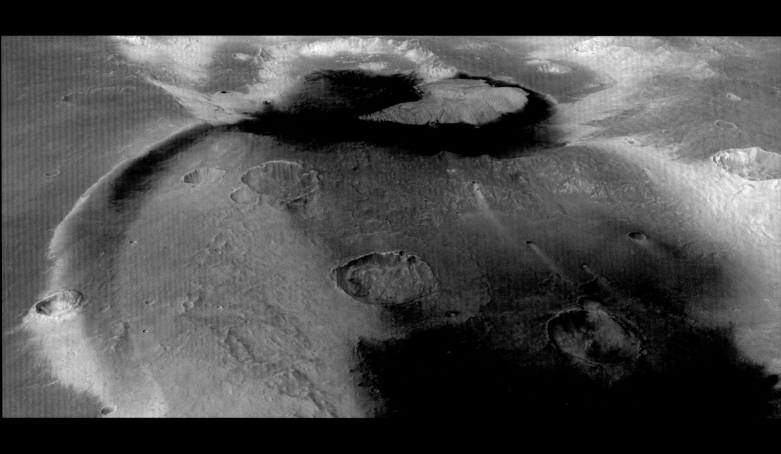

↑ A perspective view of Becquerel crater and its surroundings reveals a broad fan of dark material apparently blown southwest from the crater (in the background) and across a shallower neighboring crater (foreground). Bequerel's prominent mound of layered sedimentary rocks is prominent near the top of the picture, surrounded by dark dust with a probable volcanic origin.

→ Peering into Becquerel with its HiRISE camera, NASA's Mars Reconnaissance Orbiter reveals an astonishing landscape of exposed sedimentary layers, with dark terraces marking accumulations of basaltic (volcanic) sand. The layers themselves could have formed as either windblown or (perhaps more likely) water-carried sediments, and have subsequently been worn away by the scouring action of wind.

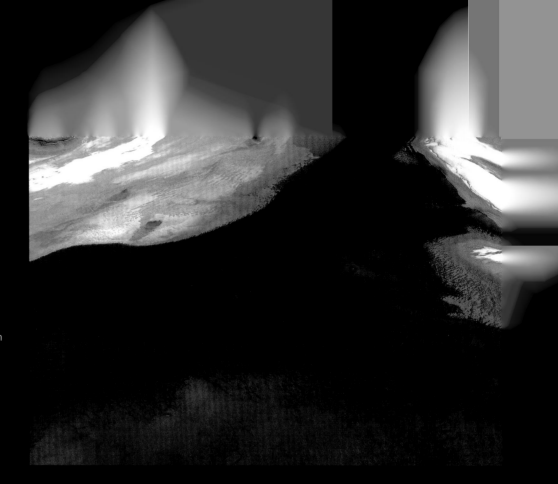

In contrast to the rugged highlands that surround the Martian south pole, the north pole is encircled by a vast and expansive plain, the Vastitas Borealis. This region, lying about 13,000 feet (4,000 meters) below the planet's average surface elevation, rings Mars at a latitude of about 70° North, with the raised polar region known as the Planum Boreum (see page 98) at its center.

Satellite-based sensors have detected the presence of large amounts of fairly pristine water ice just beneath the surface, exposed in features such as the famous "crater lake" (see page 45). Images from NASA's Phoenix lander, meanwhile, reveal a generally smooth and sandy surface, strewn with small rocks and dark polygonal features. These are thought to be a result of repeated freezing and thawing, which causes the ground to expand and contract, and resemble patterns found in Earth's own tundra regions. Elsewhere, similar processes create other patterns, such as chains of low mounds that resemble fingerprints when viewed from orbit.

The Vastitas Borealis occupies much of the Martian North Polar Basin and may once have been the floor of a great northern ocean, the Oceanus Borealis. Supporting evidence for this idea comes not only from suggestive landscape features around its edges (see page 56), but also from radar evidence that the plain's underlying rock is sedimentary in nature, as opposed to the neighboring igneous or volcanic rocks.

↑ The northern edge of Vastitas Borealis is home to some of the youngest volcanic features on Mars—lava cones up to 2,000 feet (600 meters) high such as the those shown in this Mars Express image. Lava and ash deposits around these cones are remarkably fresh, although there is no conclusive evidence for ongoing volcanic activity today.

→ This Mars Reconnaissance Orbiter view reveals undulating dunes on the floor of a crater in high northern latitudes, covered in carbon dioxide frost that is subliming back into the atmosphere during Martian spring. The dark streaks are probably dust from just beneath the surface, either exposed by landslips on steep slopes or sprayed out by geyser-like activity (see page 43).

← This HiRISE image shows the interior of an ancient crater on the Vastitas Borealis, where repeated seasonal cycles have created striking geometrical patterns. Polygonal fractures have developed through the annual expansion and contraction of ice within the soil, and tiny movements of the surface have gradually "sorted" dark boulders on the ground to follow the patterns.

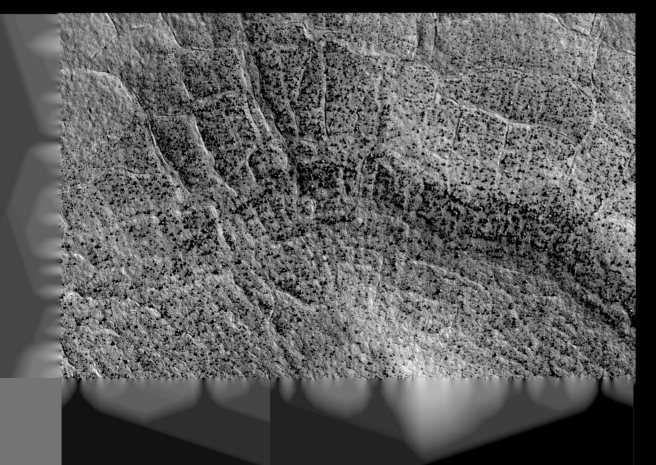

Gale crater
5.8°S, 137.8°E

Gale is a large (96-mile or 154-kilometer) crater just south of the equator in Aeolis quadrangle, a region on the boundary of the southern highlands, northern plains of Elysium Planitia and the ancient shield volcano Apollinaris Mons. It has become one of the best-studied regions of Mars thanks to its selection as landing site for NASA's Curiosity rover.

It is also one of the deepest Martian craters, with a floor about 11,800 feet (4,600 meters) beneath the Martian surface datum. Curiosity targeted the deeper, northern side of the crater and landed at the edge of an alluvial fan where water once flowed into the crater through a channel called Peace Vallis.

At the crater's center lies a peak informally named Mount Sharp by NASA mission scientists, but now officially designated Aeolis Mons. Rising up to 18,000 feet (5.5 kilometers) above the crater floor, this mountain was likely formed from the same "rebound" process that pushes up peaks at the center of some craters on the Moon. Intriguingly, the peak itself is surrounded by a huge mound of sedimentary material—layered debris whose depth suggests it piled up over around 2 billion years. Features near the top of the mound indicate that it has been sculpted by the wind, but the most popular explanation for the mound's origins is that it is the eroded remnant of deep sedimentary layers that once filled the entire crater, and elevated its floor in a similar way to that seen in the relatively nearby Gusev crater (see page 188).

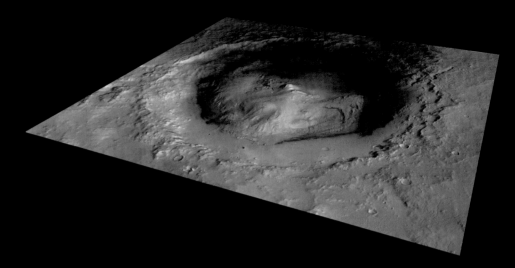

↗ This three-dimensional map of Gale shows the crater as viewed from the northwest. It was compiled using data gathered by three Mars mission—the Mars Express, the Viking orbiters, and Mars Reconnaissance Orbiter.
→ This barren, rubble-strewn view of the landscape inside Gale was captured by the 100-millimeter MastCam aboard NASA's Curiosity rover and processed by mission scientists to mimic Earthlike lighting conditions and highlight color differences between different bands in the terrain. The rising slopes of Mount Sharp dominate the background, with foothills beginning around 3.7 miles (6 kilometers) away, rising to a peak more than 10 miles (16 kilometers) distant.
→→ This colorful view from Mars Reconnaissance Orbiter combines nighttime thermal emission data (a good indicator of surface structure) with a daylit visible light image. Bluish tones, such as those seen at upper right on Mount Sharp, indicate fine-grained dust, while redder regions around the crater rim indicate harder, rocky material.

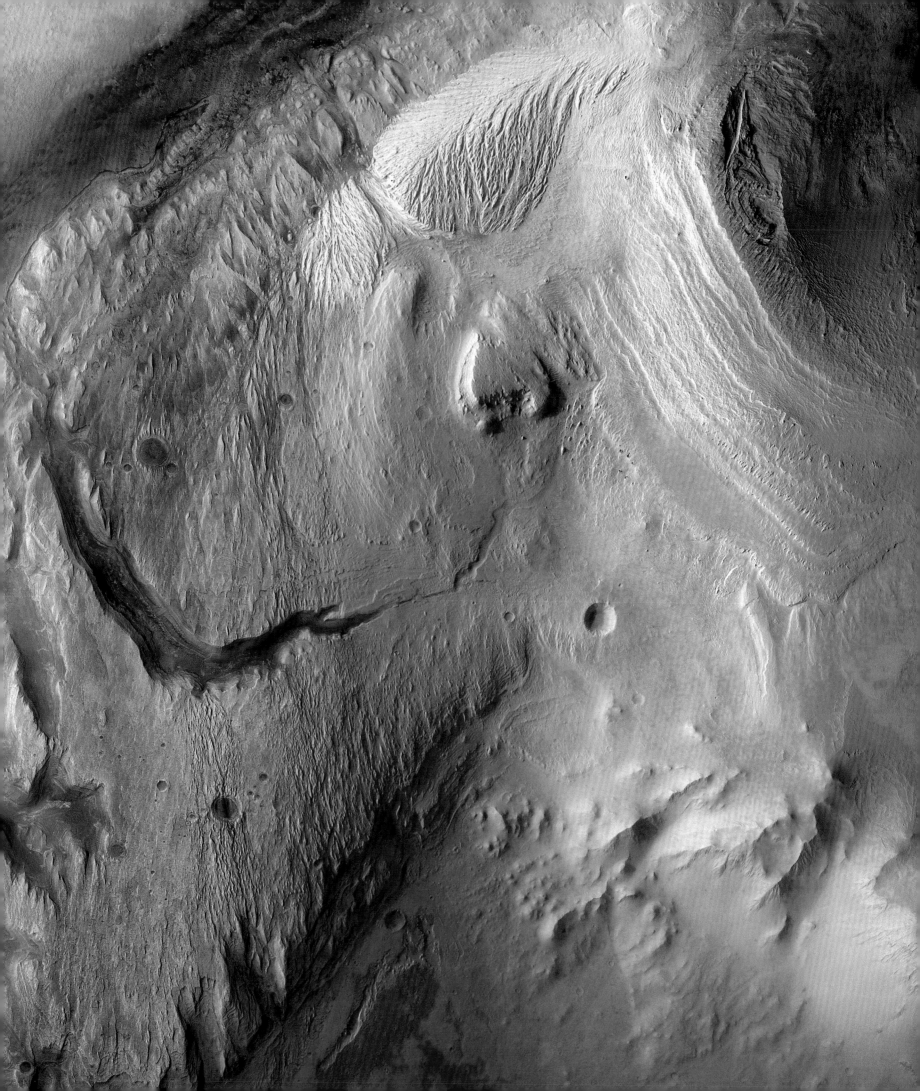

This intriguing area of the Martian surface gives its name to the distinctive unit of the crust called Arabia quadrangle. Its most striking feature is its layered surface, formed by widespread erosion of sedimentary rock formed in a previous epoch. The overall effect is very similar to that seen around the Colorado Plateau of the southwestern USA—a landscape of tablelike mesas and buttes, and craters with sharply terraced sides. In fact, aside from its overall shape, Arabia may be a far better analogue for Earth's Grand Canyon region than the more famous Valles Marineris.

Most geologists believe that Arabia Terra formed as a large impact basin early in Martian history, and many argue that it soon filled with water to create a vast lake or sea. Within the lake, sediments were slowly laid down to form new rock, while periodic climate change saw the lake evaporate and reform many times. A stop-start cycle of rock formation resulted in the development of distinctive layers. Alternative theories suggest the layers could have been laid down as dust and ice by a similar process to that which shapes the Martian poles or could even be debris from ancient volcanic eruptions (see page 32).

Within a few hundred million years, the formation of the Tharsis rise on the opposite side of Mars (see page 90) caused Arabia to push outward as a sort of counterbalance. The region's increased elevation triggered rapidly accelerated erosion, and by the time Arabia eventually subsided back under its own weight, it had formed a distinctive terrain. Since then, it has been further shaped by other processes, such as impacts and climate change.

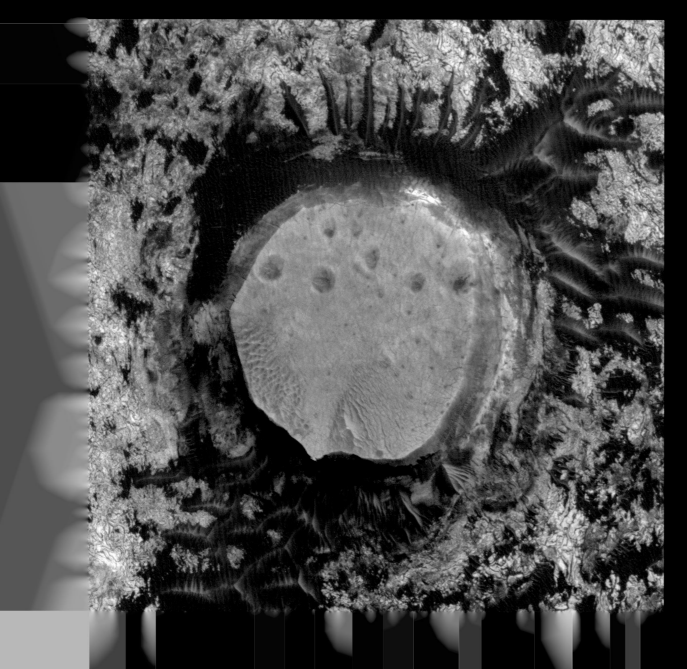

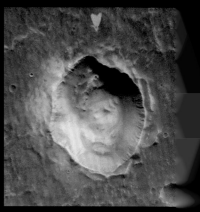

↑ The heart-shaped feature at the top of this image, roughly 0.6 miles (1 kilometer) long, was formed by a small impact whose shockwave blasted away a layer of darker material from the surface.
← This extraordinary circular plateau is an extreme example of inverted terrain—a crater with a surrounding network of channels that became entirely filled by sedimentary layers. Over time, the surrounding terrain eroded away to leave only the hardened sediments exposed.
→ This colorful view from the Mars Reconnaissance Orbiter reveals the thermal properties of a dune field trapped within a large but unnamed crater. Blue areas indicate the finest sediments, green highlights coarser-grained sands, and yellow and red are probably exposed solid rock.

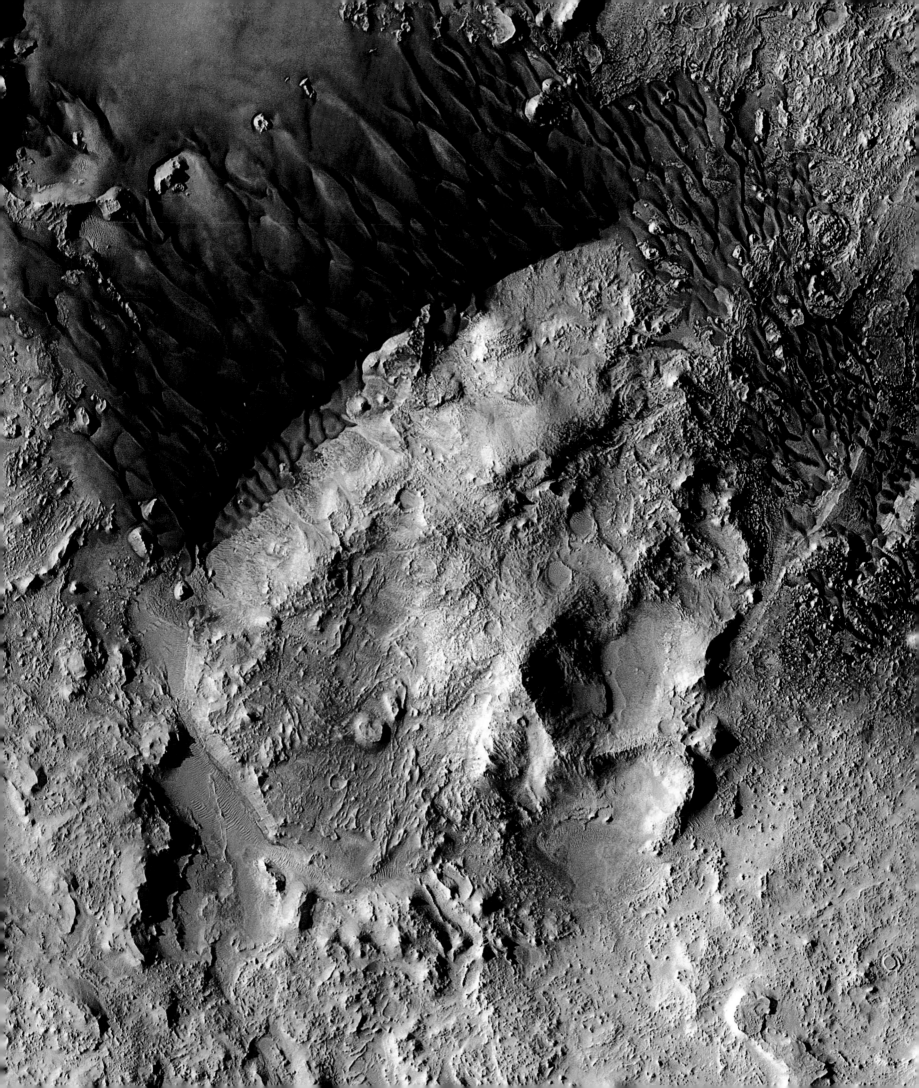

Alba Mons
40.5°N, 109.6°W

Although Olympus Mons and the Tharsis volcanoes are the most prominent volcanic peaks on Mars, they are not actually the largest in terms of area. That honor goes to Alba Mons, an enormous but low-lying volcanic shield with dimensions of roughly 1,240 x 1,860 miles (2,000 x 3,000 kilometers, or roughly the size of the continental United States), dominating the northern half of the Tharsis rise as it stretches toward the north polar plains.

Yet despite its vast extent, Alba Mons rises to a maximum of just 4.2 miles (6.8 kilometers) above the average Martian surface datum. This low relief meant that its true extent was not fully recognized until quite recently—orbital photographs revealed the enormous summit crater known as Alba Patera, but this was thought to be a curious volcanic caldera that had somehow developed without a surrounding shield. It was only in the late 1990s,

as laser altimetry data from the Mars Global Surveyor space probe began to build up a detailed three-dimensional map of the planet, that the real nature of Alba Mons became apparent.

Everything about this volcano is on an enormous scale—individual lava flows radiating out from around the central caldera complex extend for as much as 310 miles (500 kilometers), while the larger of the two overlapping calderas itself has dimensions of 105 x 62 miles (170 x 100 kilometers) and is large enough to have a smaller volcanic shield nestling within it. Long, deep parallel ditches known as fossae run for some 1,240 miles (2,000 kilometers) or more along the volcano's northwestern and southeastern flanks, and the entire landscape is covered with a deep and apparently semipermanent layer of dust up to 80 inches (2 meters) thick.

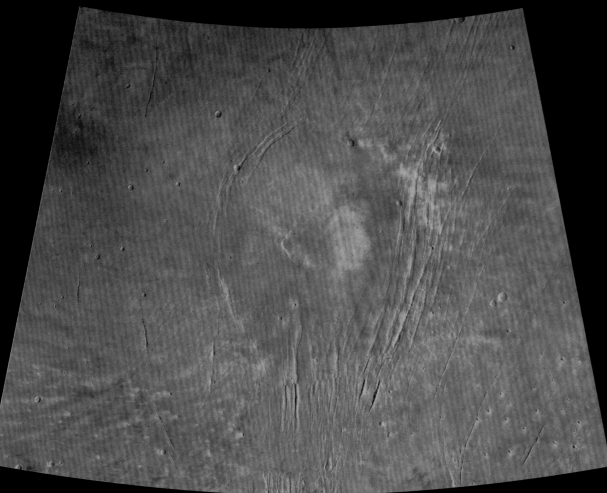

← This mosaic compiled from Viking orbiter images shows the network of fractures that surround both sides of Alba Mons. Those to the west of the central caldera (left in this image) are known as Alba Fossae, while those to the east are the Tantalus Fossae. Both sets of faults are thought to have formed due to upward pressure from the magma chamber beneath the volcano.

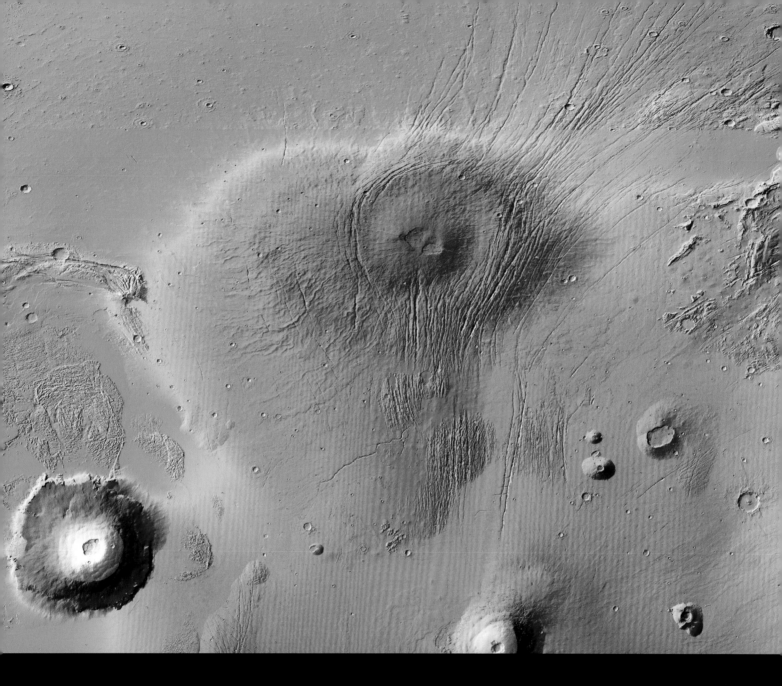

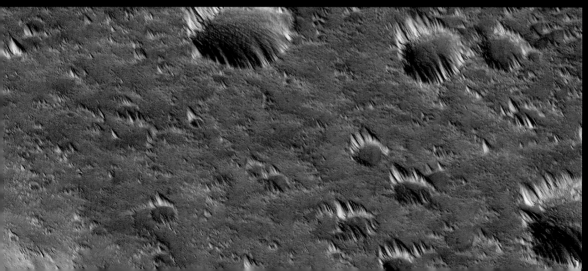

↑ This radar map from Mars Global Surveyor's MOLA instrument reveal the previously undetected shield volcano surrounding the central caldera previously known as Alba Patera, and the huge extent of the surrounding network of fossae.

← These bizarre features to the west of Alba Mons are thought to be thick deposits of dust associate with a field of impact craters. Such features, known as "whiskers," are only found in a few places on Mars and may only form in regions with the right combination of plentiful dust and a low elevation providing a relatively thick atmosphere.

With an evocative name that translates as "Labyrinth of Night," the Noctis Labyrinthus lies at the western end of the Valles Marineris (see page 84) and bridges the gap between that enormous canyon system and the uplands of the Tharsis rise to its northwest. Its mazelike landscape consists of countless blocky, flat-topped mesas separated by deep, steep-sided valleys. Many of the valley floors are flat themselves, with features that suggest they are "grabens"—a type of geological fault in which sections of the landscape sink vertically below their neighbors while remaining otherwise undisturbed. Others, however, are littered with debris from landslides down the canyon walls.

Most experts believe that the labyrinth formed as a result of widespread subsidence at the edge of the Tharsis rise. Volcanic activity, the theory suggests, heated subsurface ice and water in the nearby rocks, causing it to escape into the Martian atmosphere and undermining the landscape from within, until eventually certain regions subsided. The rock beneath the mesas was presumably not undermined in the same way, so we might even interpret the mazelike geography of the present-day Noctis Labyrinthus as an echo of ancient subterranean watercourses.

Recently, the labyrinth has provided another window into the Martian past through the study of deposits on its valley floors. So-called "light-toned deposits" (LTDs), formed by landslips as material came away from the valley walls, show telltale signs of complex minerals such as sulfates and clays that are likely to have formed in the presence of water.

→ This colorful image from Mars Odyssey's THEMIS infrared camera highlights temperature differences. Loose rubble from a huge landslide, perhaps triggered by the formation of a 3.7-mile (6-kilometer) crater on the valley rim, spreads out in a broad fan across its floor, appears red because it is relatively warm.

↓ At the western end of the Valles Marineris, sharply defined canyons give way to a jumbled, mazelike landscape of isolated mesas with narrow valleys between them. The dune-filled valley floor between two such mesas is shown in stunning detail overleaf.

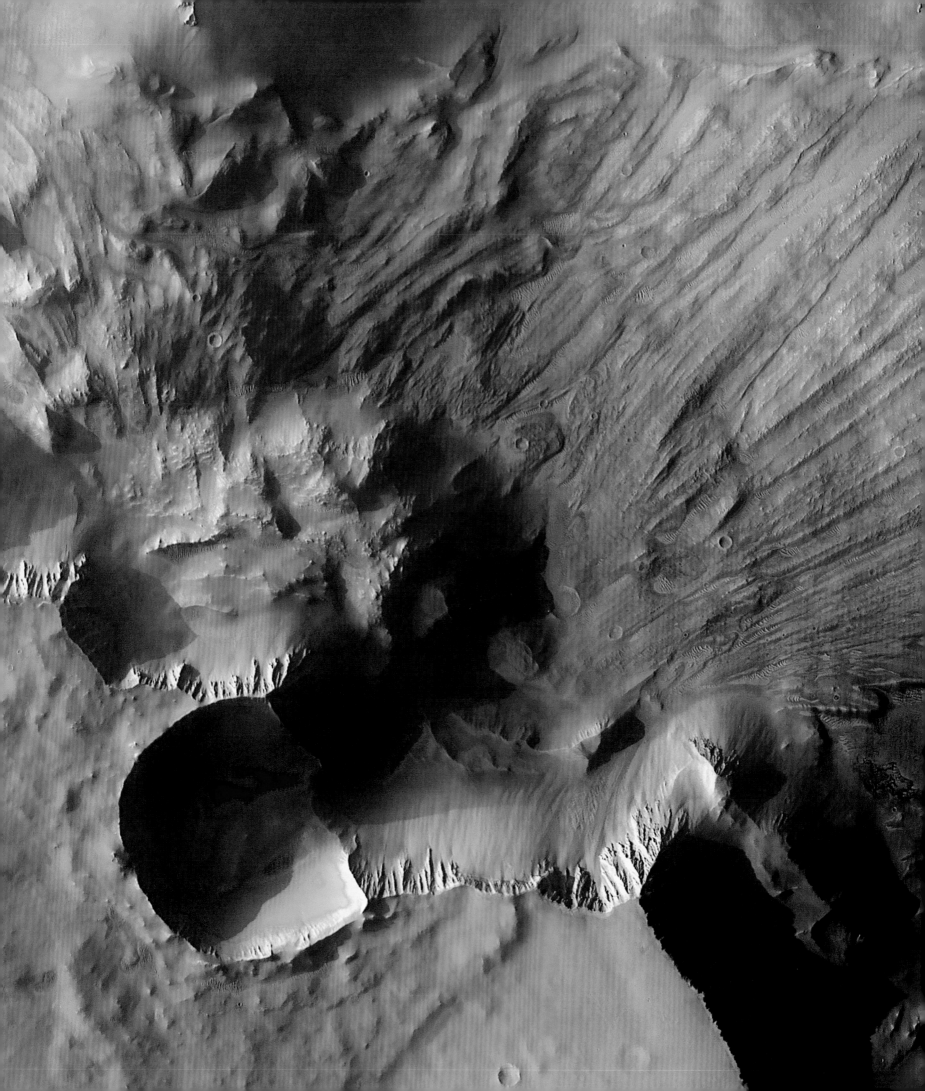

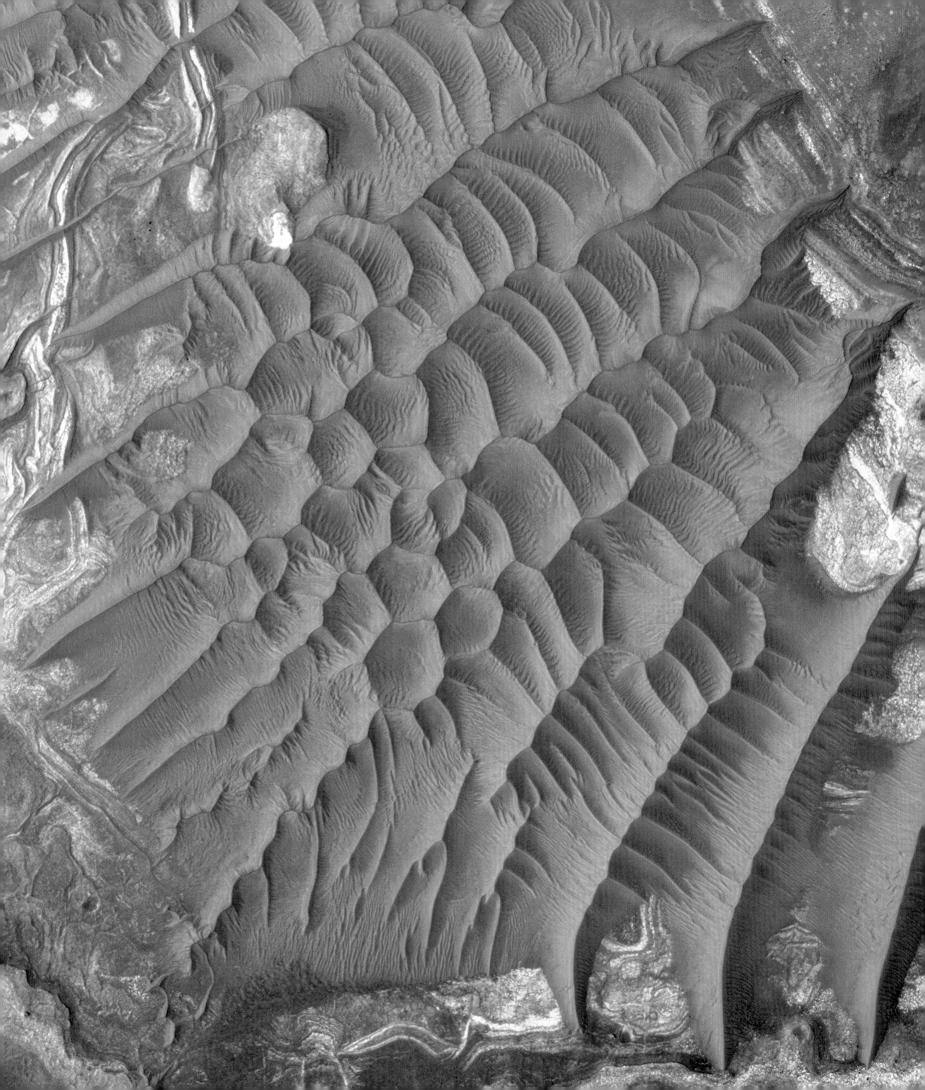

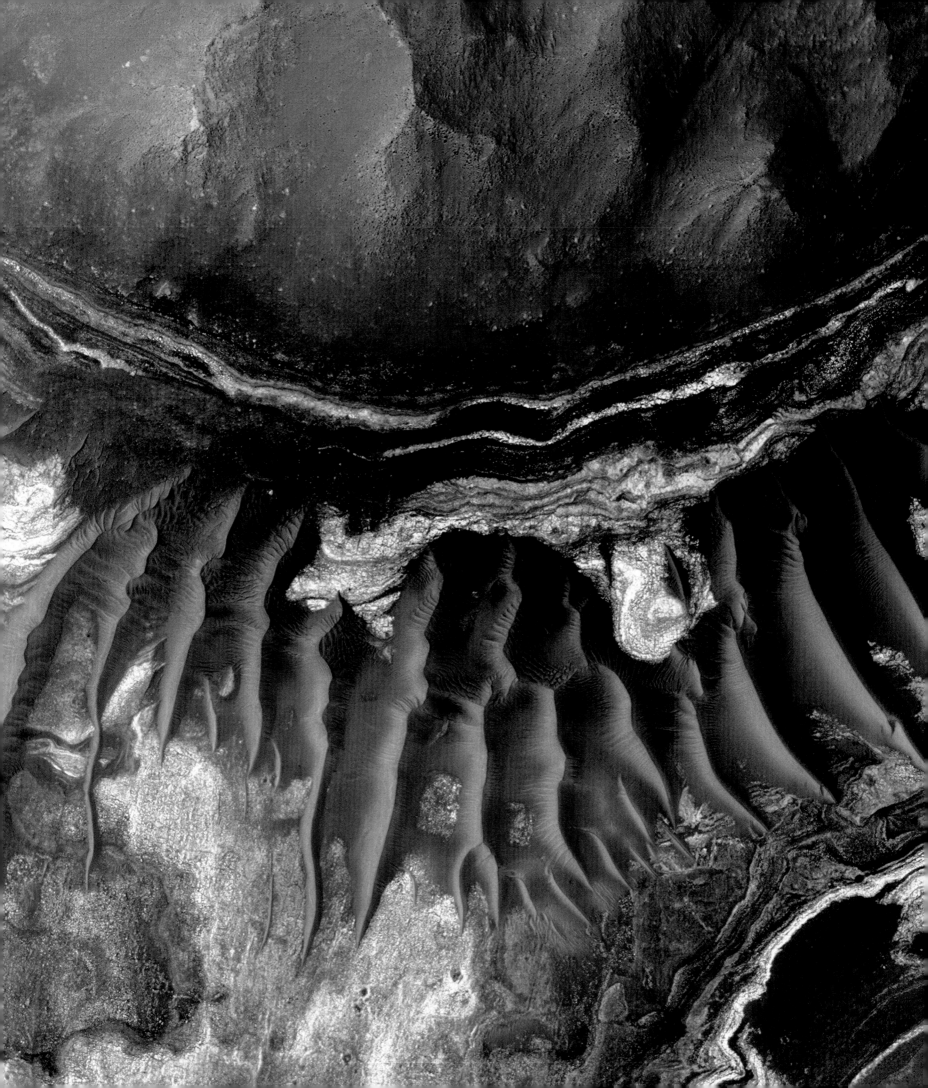

Ismenius Lacus

47.5°N, 30°E

Although its name suggests a reference to a specific lakelike feature, Ismenius Lacus is today defined as a broad region or quadrangle of the northern hemisphere, with a name inherited from an earlier dark albedo feature mapped by 19th-century astronomers. It marks the northernmost extension of the boundary between southern highlands and northern plains, and latitudes are high enough for ice to have played a significant role in shaping the landscape. As a result, it plays host to some unique landscape features along the transition zone.

Broadly known as fretted terrain, these regions are characterized by a gradual change from the relatively narrow, straight-sided valleys that run through the highlands into isolated mesas on a lowland plain—features that geologists on Earth would attribute to the action of ancient glaciers. What's more, the mesas are often surrounded by so-called "lobate debris aprons," which radar investigations have shown are mostly composed of ice with a thin layer of overlying rocks. The ice that shaped this landscape is still at work today.

To the north, Ismenius Lacus quadrangle extends to cover a substantial area of the Vastitas Borealis northern plain (see page 110). Here, it encompasses the 147-mile (236-kilometer) Lyot crater, one of the most prominent Martian craters with a floor that is the deepest point in the northern hemisphere.

→ The Ismenius Lacus quadrangle contains two major areas of fretted terrain—Protonilus Mensae and Deuteronilus Mensae (part of which is shown here). This Mars Express image is dominated by a broad (68-mile/110-kilometer) depression filled with dark material. The region is believed to have been carved out by glacial meltwaters flowing down from neighboring highlands.
↓ A Viking orbiter mosaic of Ismenius Lacus quadrangle shows how the landscape varies from flat lowland plains in the north, through fretted terrain to elevated highlands. Lyot crater sits close to the center of the quadrangle.

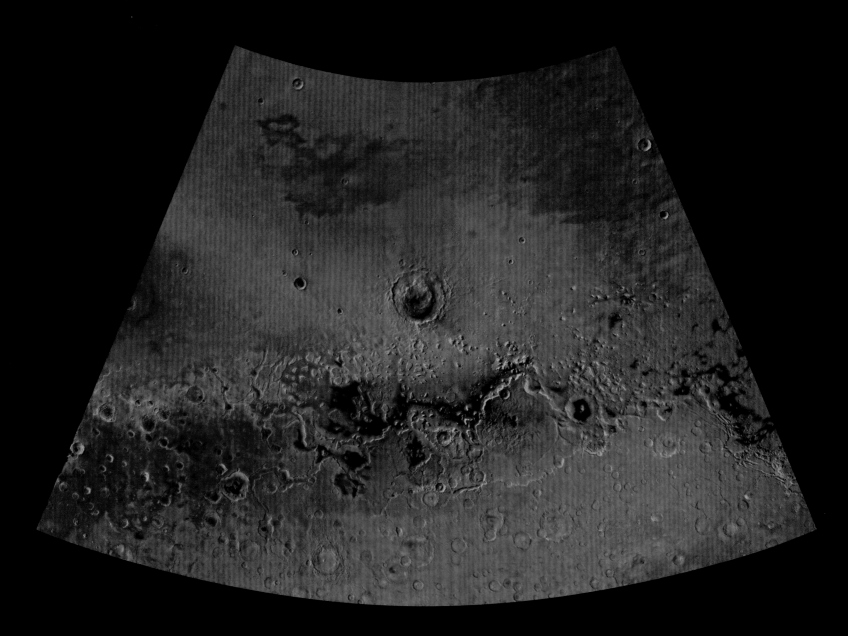

0°

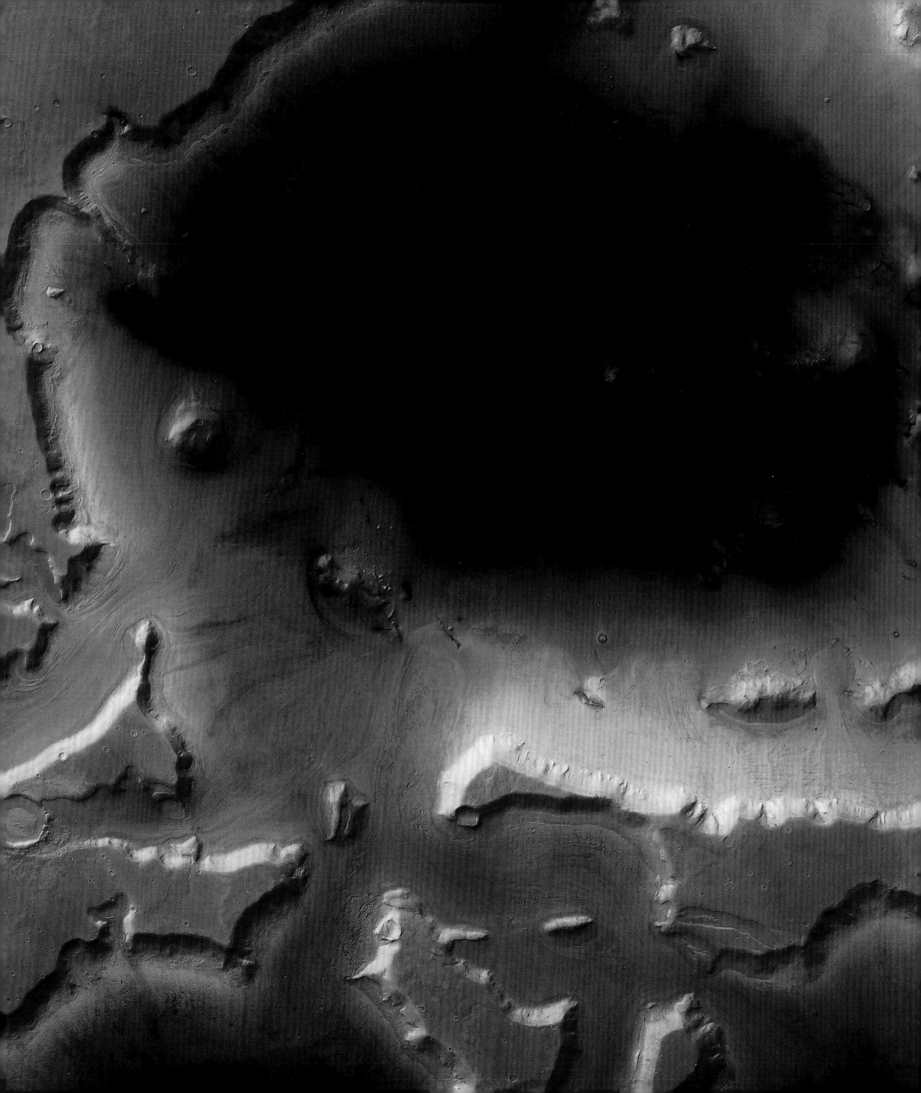

The Cydonia region lies in the northern hemisphere of Mars, where the southern highlands descend toward the northern plains. It is a region of flat-topped mesas, hummocky and knoblike hills, and intersecting valleys, that takes its name from a feature on preSpace Age maps, itself named after an ancient city-state on the Mediterranean island of Crete. During the wet early phase of Martian history, Cydonia's location would have rendered it vulnerable to flooding and erosion as water flowed down from the highlands toward low-lying lakes and the hypothetical Oceanus Borealis, and this was probably responsible for shaping most of the landscape features seen today.

However, Cydonia is perhaps best known as the location of curious features captured in early Viking space probe images, which some excitable observers interpreted as ruins of an ancient Martian civilization. Among pyramids and polygonal hills, the most famous of these is the so-called "Face on Mars," a 1.2-mile (2-kilometer) oblong mesa with a striking resemblance to a human face. Although NASA scientists immediately recognized the face as a "trick of light and shadow," this did little to discourage conspiracy theorists or more general speculation. Cydonia has since been revisited by both the Mars Global Surveyor and Mars Express probes, both of which have returned more detailed images and even three-dimensional views confirming that the face is indeed an illusion, while the other apparently geometrical features have natural origins.

→ This famous image captured by Viking orbiter 1 in July 1976 sparked more than two decades of debate about allegedly artificial features in the Cydonia region, most famous of which was the apparent "Face on Mars."

→→ A Mars Express image of the Cydonia region, captured 30 years after Viking's famous image, reveals a variety of features reminiscent of chaos terrain elsewhere on the planet. The "face" is visible just to the right of center, with another suggestive-looking mesa, often compared to a skull, just below it.

↓ A three-dimensional image of the so-called "face," derived from Mars Express observations, reveals that it is nothing more than a highly deformed mesa that takes on a misleading appearance in a certain light. The mesa is surrounded by an apron of eroded debris, while a huge landslip is clearly visible on its near side.

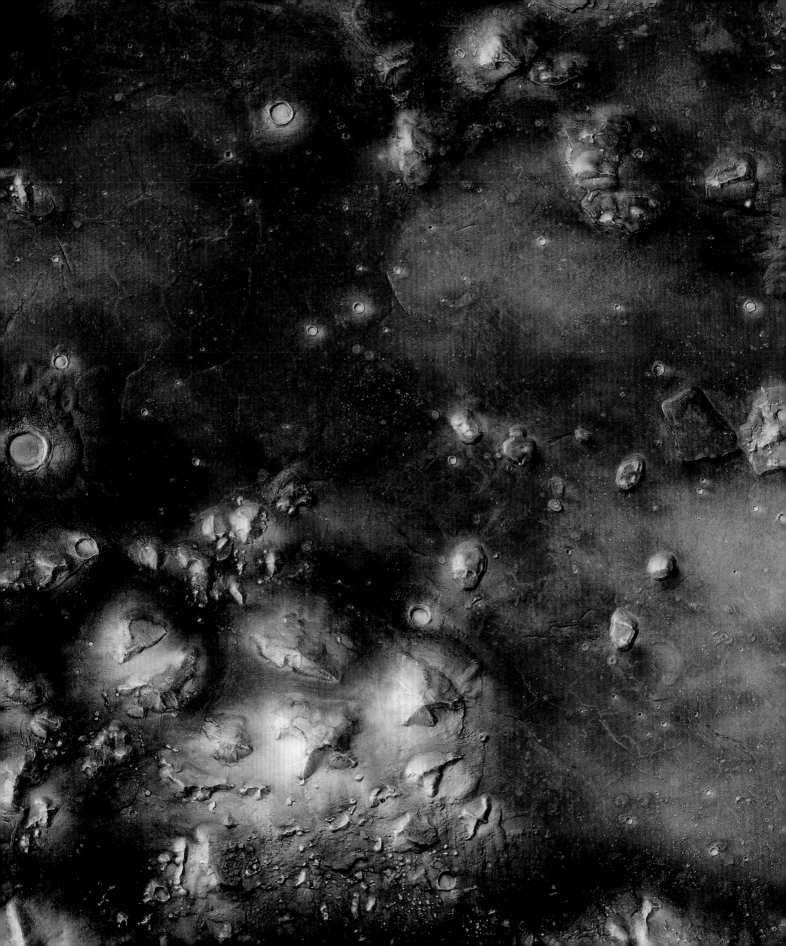

has subsequently gone through several stages of transformation linked to its location bordering both Tharsis and the southern highlands. Within a few hundred million years of its formation, it seems to have been flooded by volcanic lava that created a mostly smooth basin floor. As the lava set into a solid crust with a still-molten layer beneath, it piled up in places to form distinctive "wrinkle ridges"

a semi-permanent Chryse lake (or a bay in the hypothetical great northern ocean). At a later stage, as the Martian climate became cooler and drier, Chryse was the collecting point for torrential waters released in catastrophic floods that scarred the landscape to create distinctive "outflow channels" such as Kasei Valles and Ares Vallis (see pages 136 and 156).

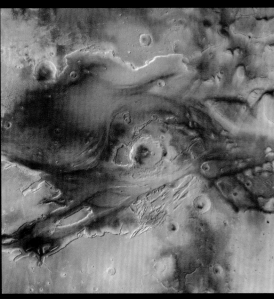

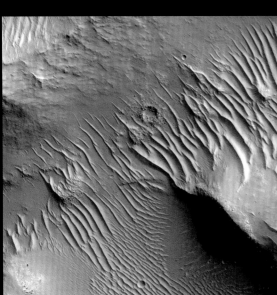

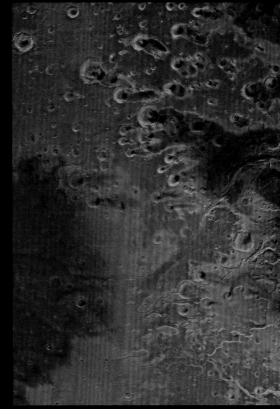

← This Viking orbiter mosaic focuses on the complex outflow region that drained the southern highlands into Chryse Planitia near the top of the picture. The broad valleys, formed in catastrophic floods, cut up to 0.6 miles (1 kilometer) deep through the surrounding landscape.

←← A Mars Express image shows further outflow channels and isolated islands of raised terrain to the west of Chryse, around the mouth of Kasei Valles (see page 136).

↙ Delicate dunes cover the floor of the upper Samara Valles, one of the southern outflow channels that drain into Chryse.

→ A close-up view from NASA's Mars Reconnaissance Orbiter shows a jumbled region of small mounds set amid the surrounding scablands of Chryse Planitia.

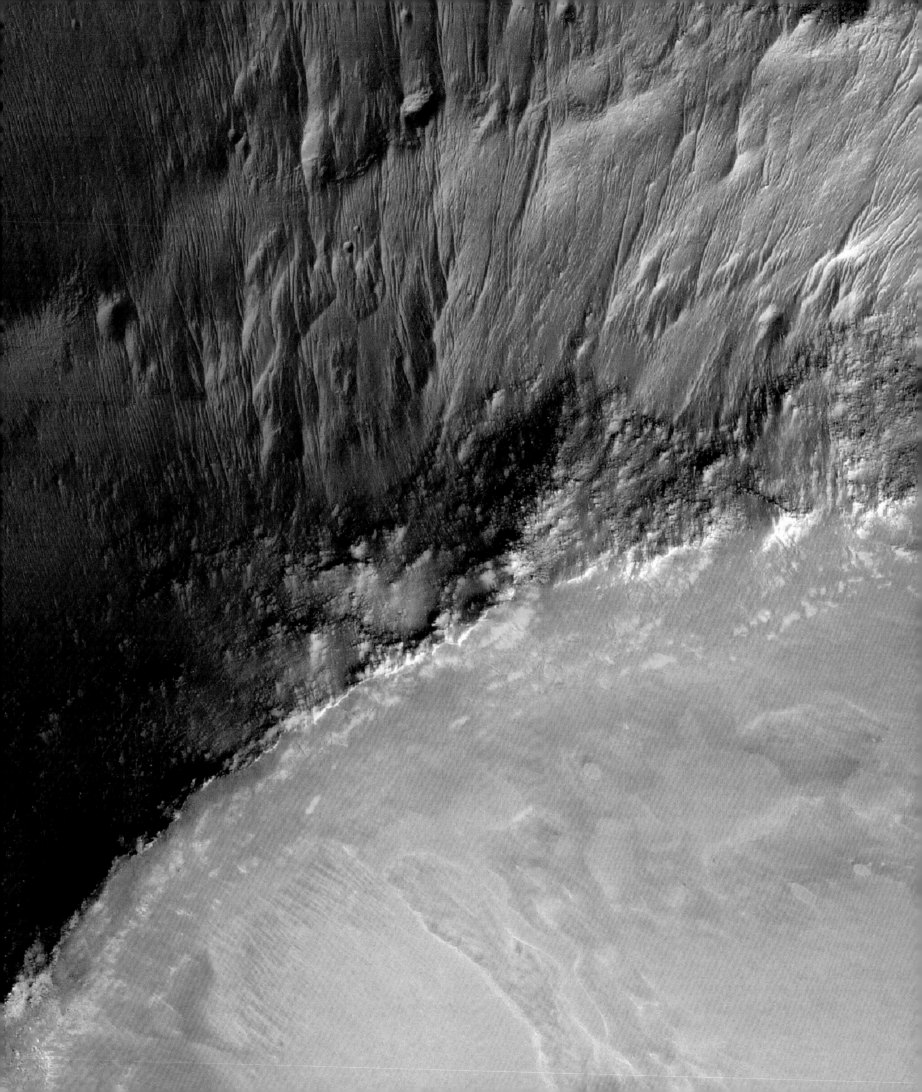

Victoria is informally named after the first ship to circumnavigate the world, as part of the Portuguese explorer Ferdinand Magellan's expedition of 1519–22. Some 2,400 feet (730 meters) in diameter, its most distinctive feature is its scalloped edge, created as soft material on the crater rim has eroded and collapsed. Opportunity traveled roughly a quarter of the way around the

After identifying a suitably shallow and stable slope near the alcove known as Duck Bay, Opportunity descended into the crater in September 2007 and subsequently spent almost a whole year making its way around Victoria's inner wall, paying particular attention to the exposed sedimentary rocks exposed on the face of a promontory known as Cape Verde.

→ Mars Reconnaissance Orbiter captured this image of Victoria crater in October 2006. Duck Bay, Opportunity's entry point to the crater floor, is at the 10 o'clock position, with Cape Verde just above it. Opportunity itself can be seen as a tiny dark spot just on the crater lip at Cape Verde.

← Victoria's central field of dunes covers an area of roughly 1,120 x 790 feet (340 x 240 meters). Its extraordinarily intricate patterns are thought to be a result of interaction between steady and persistent winds blowing from different directions across the crater floor.

↓ This panoramic view of the crater was taken by Opportunity as it stood on Cape Verde. The outcrop at left is known as Cape St. Mary, while the smooth slopes of Duck Bay are to the right.

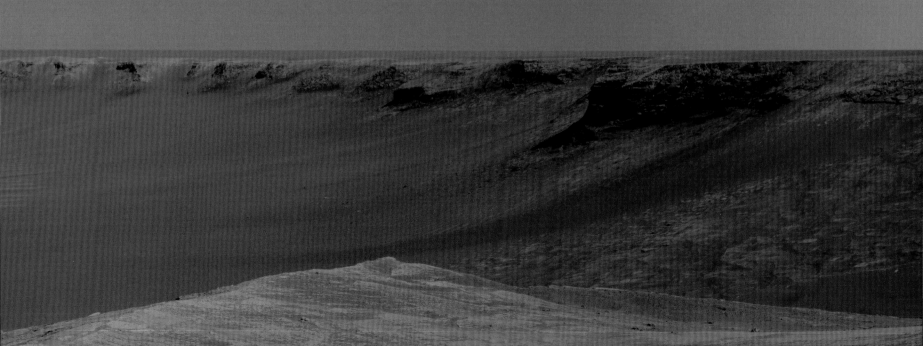

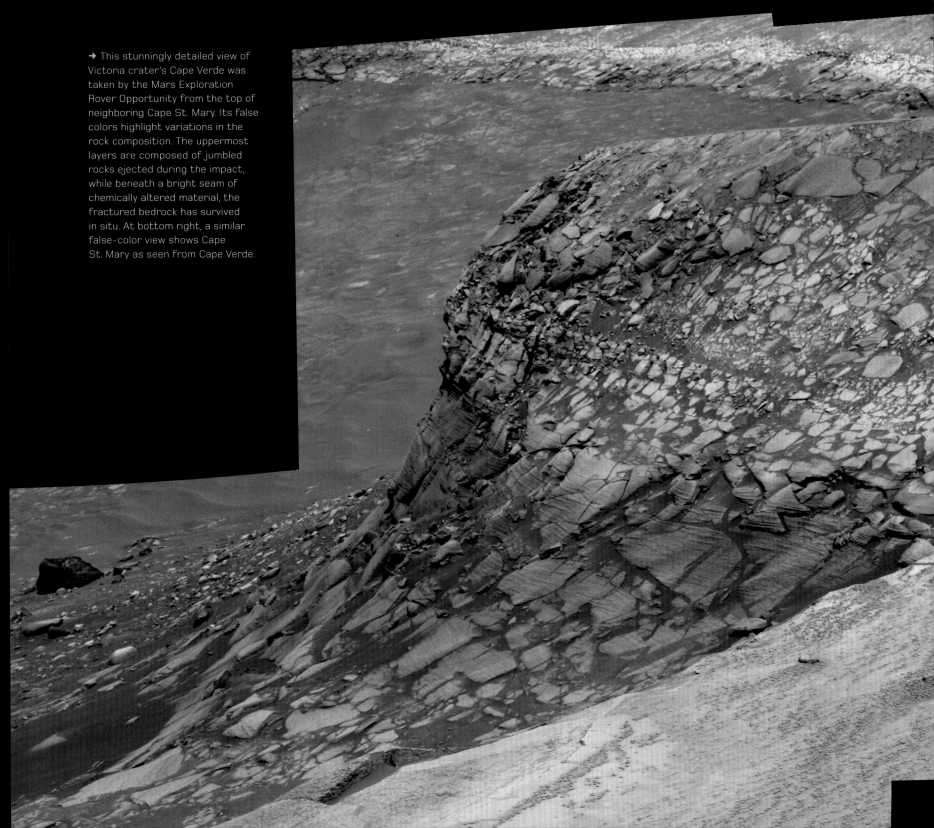

→ This stunningly detailed view of Victoria crater's Cape Verde was taken by the Mars Exploration Rover Opportunity from the top of neighboring Cape St. Mary. Its false colors highlight variations in the rock composition. The uppermost layers are composed of jumbled rocks ejected during the impact, while beneath a bright seam of chemically altered material, the fractured bedrock has survived in situ. At bottom right, a similar false-color view shows Cape St. Mary as seen from Cape Verde.

Noachis Terra
45°S, 10°W

The highland region of Noachis Terra (whose name translates as "Land of Noah") lies to the west of the enormous Hellas impact basin. Though heavily cratered, it is best known for the spectacular dune fields found within its individual craters.

Apparently composed of several different grades of sand, these dunes are sculpted into impressive structures, some of which mirror patterns seen on Earth, while others seem entirely unique to Mars. The shapes of the dunes are probably linked to prevailing wind patterns within individual craters, which are themselves linked to local changes in temperature throughout each Martian day.

Detailed studies of the region suggest that the sand now forming dunes mostly originated within the separate craters, rather than being supplied from more remote sources or simply dumped by global dust storms. It seems likely to have eroded out of distinctively layered sedimentary material that was already in place by the time the craters formed. Given Noachis Terra's upland location, any sediments are unlikely to have been laid down in standing bodies of water, so instead, the most likely sources of the original sediment are volcanic ashfall or the action of glaciers. Regardless of the precise details of the mechanism at work, its results are little short of astonishing.

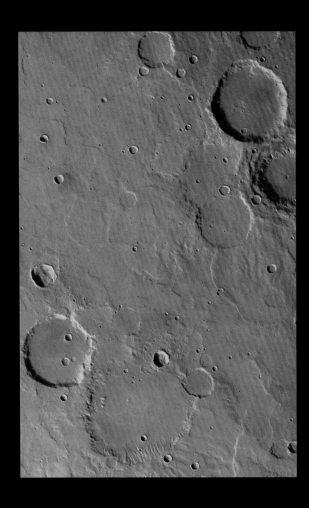

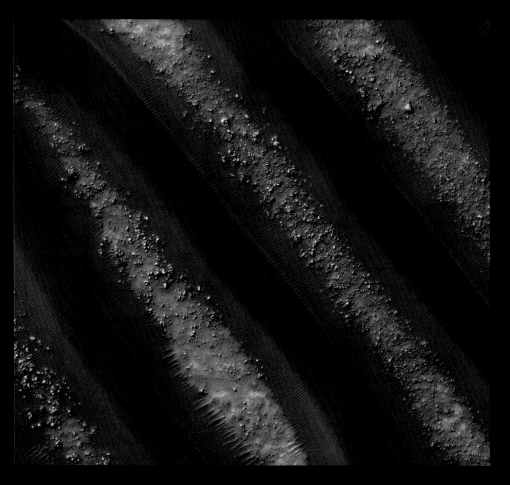

↑ This Mars Express image shows an extensive region of northern Noachis Terra, covering an area of roughly 35,000 square miles (91,000 square kilometers). Broad, multilayered lava plains can clearly be seen—a feature that defines the entire Noachian period of Martian history (see page 26).

↑ This remarkable image from Mars Reconnaissance Orbiter shows the power of Martian winds to sculpt geometric structures over millions of years. These linear dunes, spotted inside a Noachis Terra crater, create almost perfect stripes across the crater's rubble-strewn floor.

→ Thick dust provides an ideal medium for the artistic power of wind across Noachis Terra— this HiRISE camera image shows extraordinarily beautiful dune patterns forged from the sand in another highland crater. The image resolves objects down to just 30 inches (76 centimeters) across.

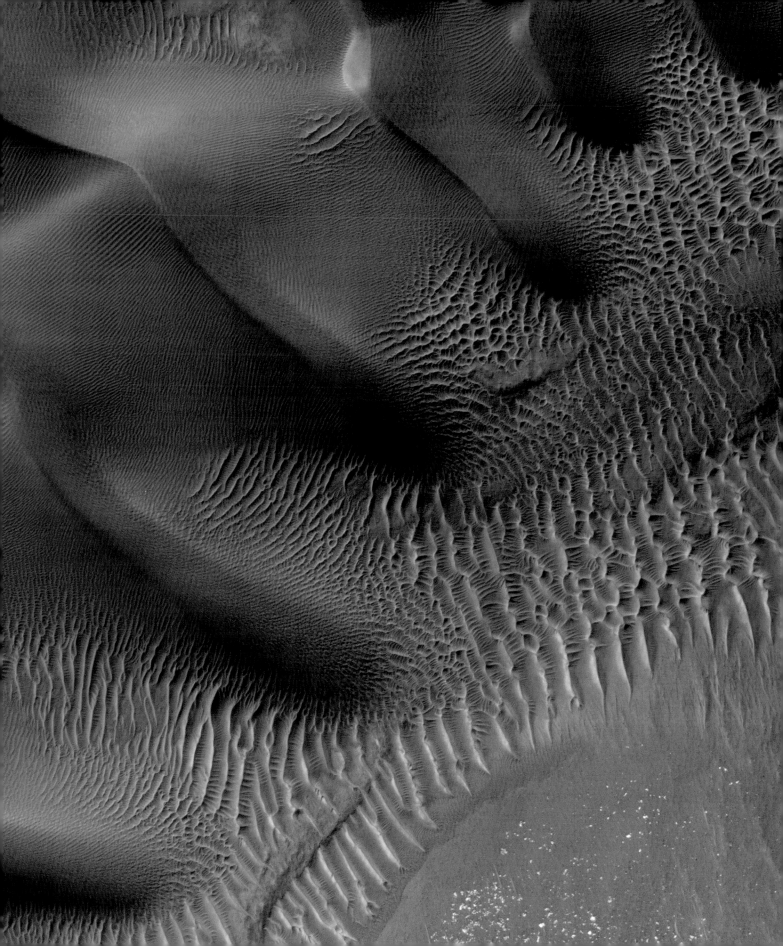

Elysium Planitia
3.0 N, 154.7°E

Halfway around the planet from the Tharsis rise, Elysium is the second largest volcanic province on Mars, dominated by the huge shield volcano Elysium Mons and its smaller companions Hecates Tholus and Albor Tholus. To the south of the volcanic rise lies a broad plain of intermediate altitude known as Elysium Planitia, which shows many signs of being shaped by the action of water in the surprisingly recent past.

Images from the HRSC camera on ESA's Mars Express space probe revealed in 2004 the presence of what appears to be a large "frozen sea," roughly 500 x 560 miles (800 x 900 kilometers) across. Its surface is broken into several large plates covered by a thin layer of volcanic ash, while its average depth is estimated to be around 150 feet (45 meters). Most remarkably, the paucity of impact craters on its surface, and its relationship to other features around it, suggests that this "sea" may be just a few million years old. A leading explanation is that pressure from molten magma beneath Elysium created a nearby series of parallel faults known as Cerberus Fossae, allowing water to pour out from beneath the surface in a catastrophic flood that carved the Athabasca Valles canyon complex as it forced its way down into Elysium.

Some skeptics point to the fact that the few impact craters in this area do not seem to have exposed the expected pristine ice beneath, and have even argued for other mechanisms that could have formed similar landscape features. Nevertheless, the frozen sea of Elysium is certainly one of the planet's most intriguing features and, for this reason, the area has been targeted for exploration by NASA's forthcoming InSight mission (see page 210).

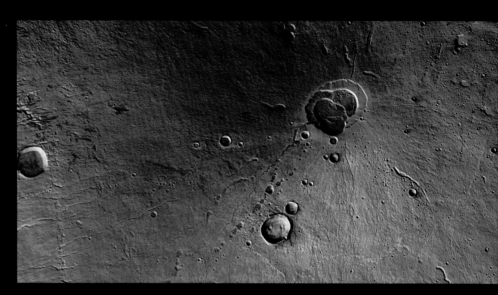

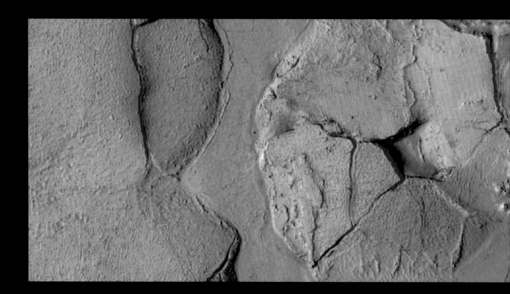

↑↑ The northernmost of three major volcanoes in the Elysium region, Hecates Tholus is a dome-shaped mountain 113 miles (182 kilometers) in diameter. Its central caldera, shown here, is around 6 miles (10 kilometers) across. The volcano is thought to have experienced its last major eruption 350 million years ago.

↑ Fractured mounds along the southern edge of Elysium Planitia, shown in this stereoscopic view from Mars Reconnaissance Orbiter, are thought to have been pushed up from beneath by tectonic forces. This image appears in three dimensions when viewed through red-blue "3D glasses."

→ Another close-up view of the

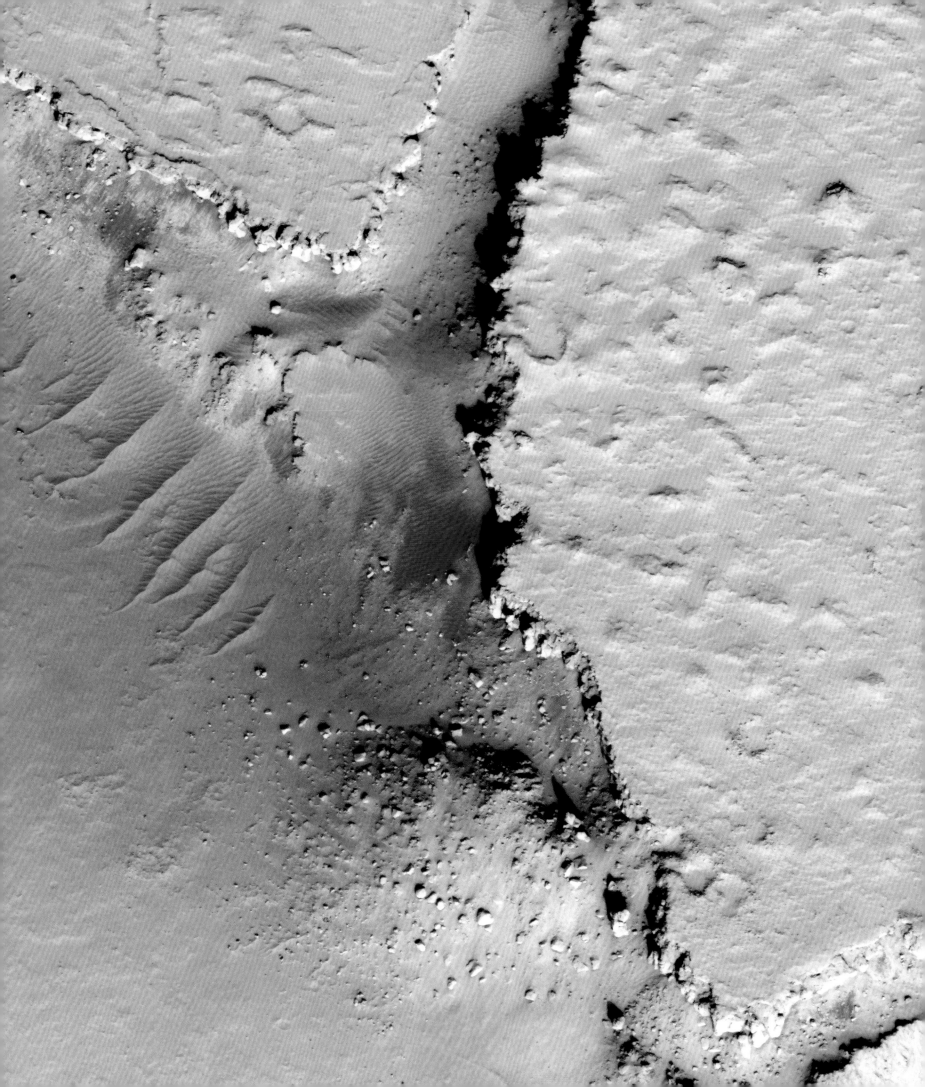

Kasei Vallis
24.6°N, 65.0°W

Named in reference to the Japanese name for Mars, Kasei Valles are a huge complex of canyons almost 1,100 miles (1,800 kilometers) long, originating close to the Echus Chasma region of the Valles Marineris (see page 84) and emptying out into Chryse Planitia. At their broadest, these valleys are up to 310 miles (500 kilometers) wide, branching into multiple channels that isolate large, flat-topped islands known as mensae (from the Latin for "table").

Kasei seems to mark the path taken by one or more huge and violent floods, most likely of liquid water escaping from beneath the Martian surface around 3 billion years ago in the Hesperian or early Amazonian periods. The floods themselves may have been triggered when volcanic heat from the nearby Tharsis rise melted underground ice, and would have drained into Chryse Planitia and perhaps beyond into the hypothetical Oceanus Borealis surrounding the Martian north pole.

As the water cascaded across the landscape, it left teardrop-shaped scars around the hardest outcrops and in other places created plunging waterfalls that can still be traced as clifflike cataracts up to 1,300 feet (400 meters) high. Elsewhere, rocks carried on the floods have left rolling trails across the landscape.

↓ This Mars Express image shows chaotic terrain along the boundary between the Kasei Valles flood channel (near the top) and a highland plateau called Lunae Planum. The eroded crater on the right is about 22 miles (35 kilometers) in diameter.

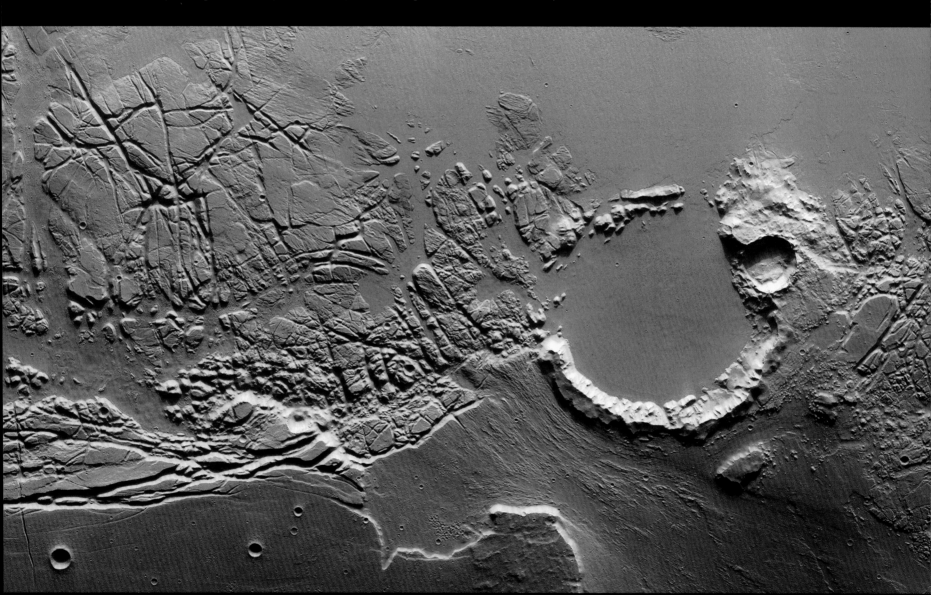

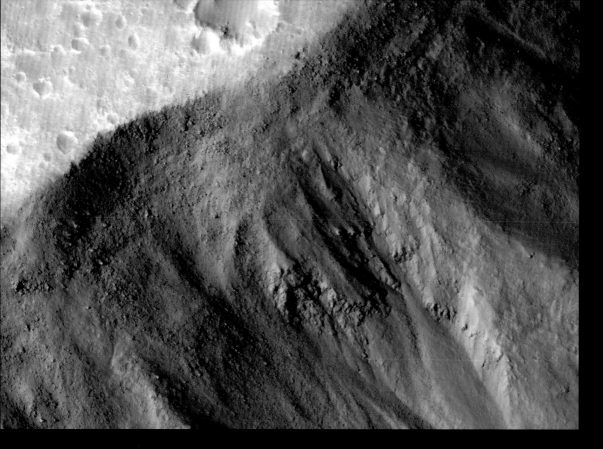

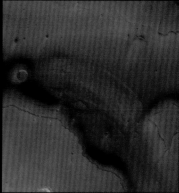

↑ Dark material around the mouth of Kasei Valles mark accumulations of basalt-rich dust on the valley floor, while bright streaks indicate "wind shadows"—areas protected from the prevailing winds, where dust was not deposited in such large amounts.

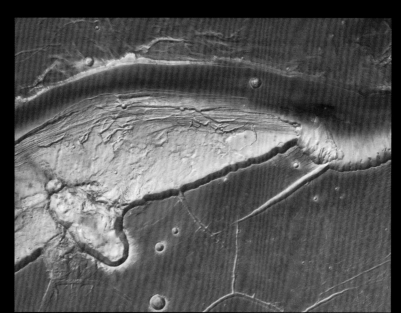

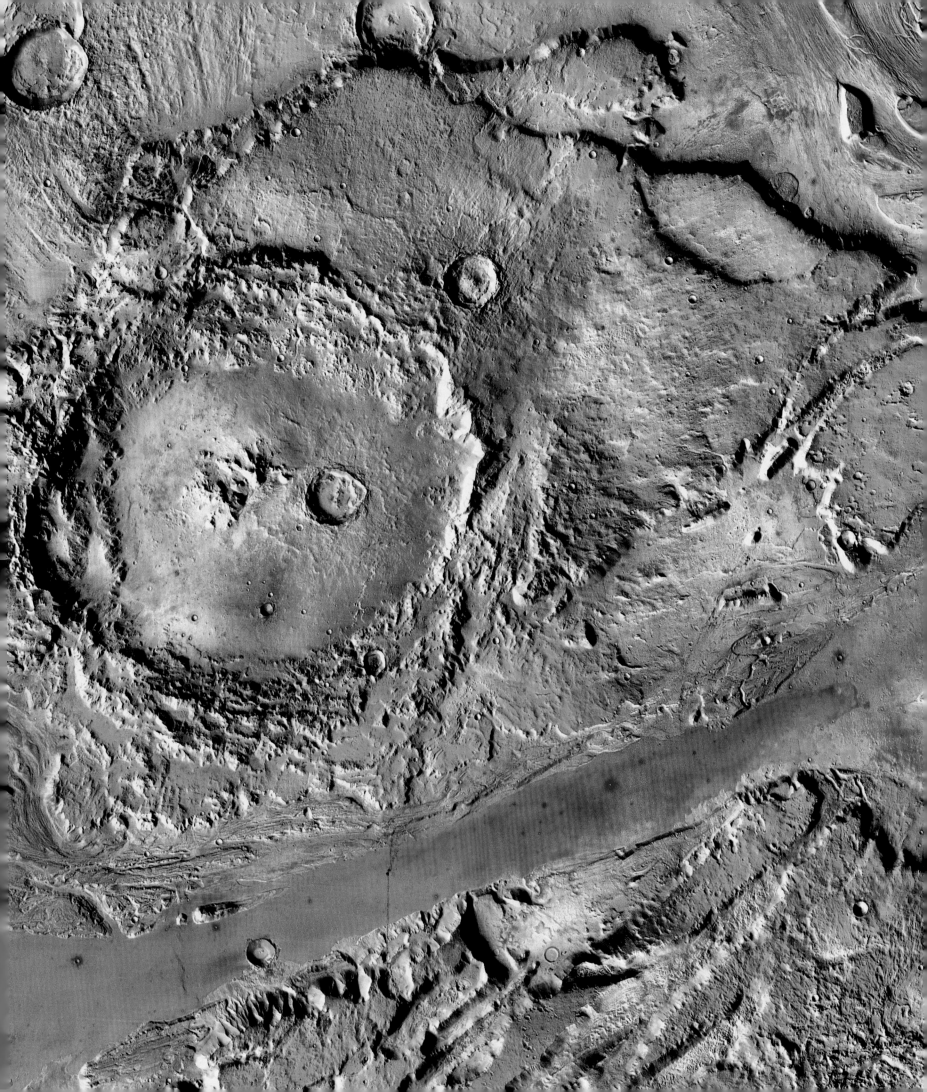

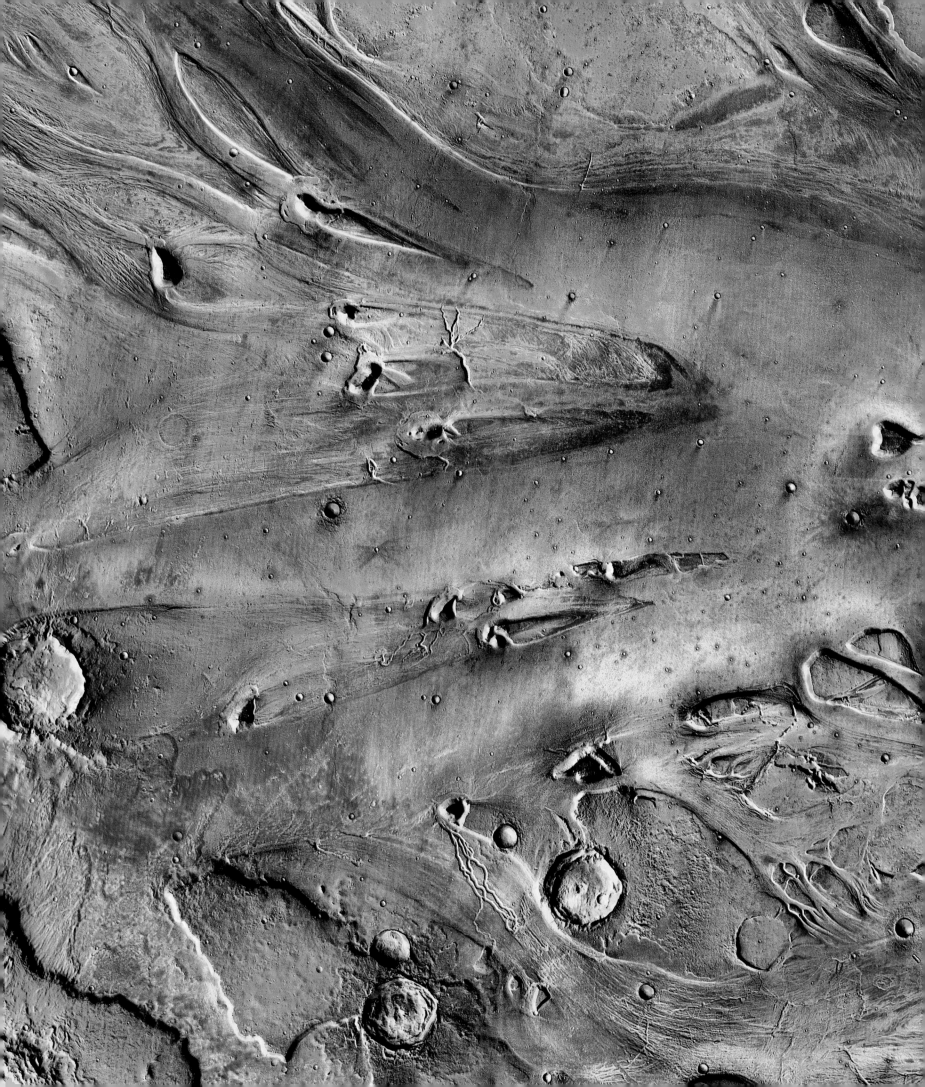

Aureum Chaos
4.4°S, 26.5°W

With a name derived from the Latin word for gold, Aureum Chaos takes its name from an albedo feature noted on early 19th century maps of Mars. The word "chaos," meanwhile, indicates a region where the surface has subsided with spectacular results. The result is a network of canyons, mesas, sunken blocks, and rolling, disorderly hummocks and knobs, covering more than 46,000 square miles (120,000 square kilometers) of the Margaritifer Sinus quadrangle.

Aureum Chaos lies at the northeastern end of the Valles Marineris, south of Chryse Planitia. Like other chaos regions, it is almost certainly the result of a dramatic removal of water or ice from the Martian soil. Satellite mapping of the area's mineral composition has suggested the presence of clays and other hydrated minerals, reinforcing the idea that water was once plentiful in these sedimentary rocks. While the subsidence that shaped this distinctive landscape formed in the Late Hesperian, perhaps 2,500 million years ago, there is unexpected evidence for more recent activity here. Detailed studies of landslides in the area that are just a few million years old have also suggested that faulting processes may be at work in the region today, pushing blocks of Martian crust past each other just as they seem to be in the Valles Marineris themselves (see page 84).

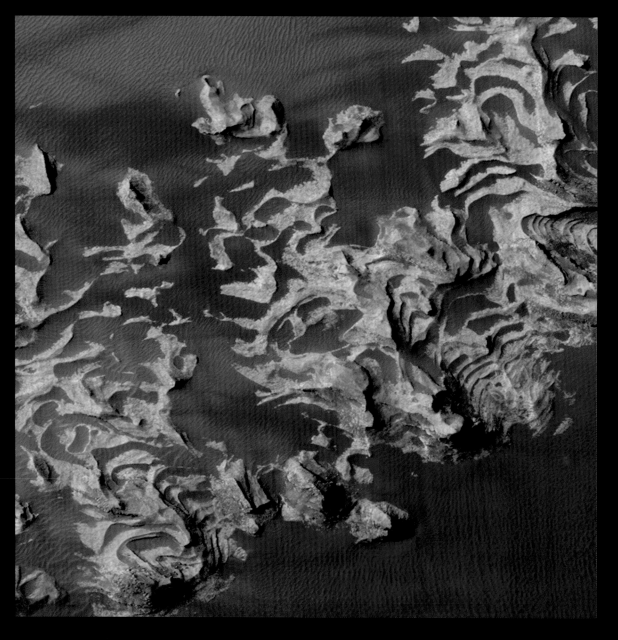

← Half-buried by later sands, this rocky outcrop shows a complex layered structure. Researchers suspect it may have built up through the accumulation of sediments after the initial collapse of the Aureum Chaos region, only to be eroded away again at a later stage.

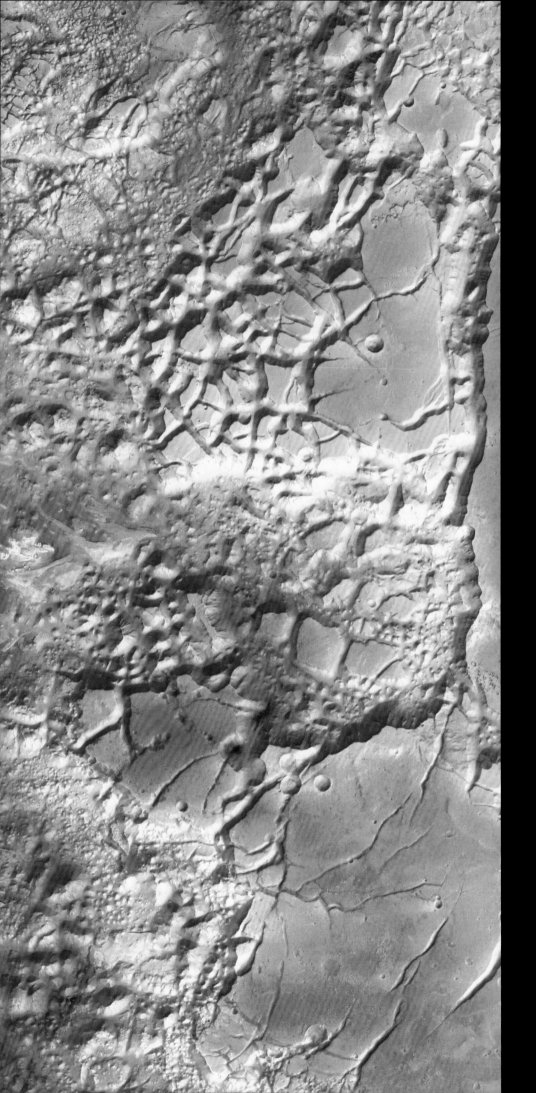

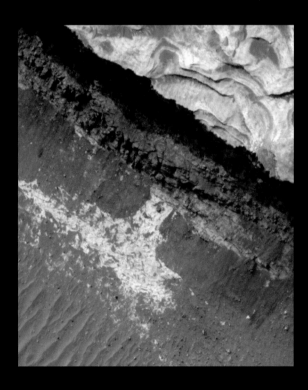

↑ A close-up view from Mars Reconnaissance Orbiter focuses on the edge of a single mesa. The top of the mesa, at upper right, is a complex swirl of layered sediments, while the valley floor beneath reveals distinct yellow and white rock units of different ages.

← A wide-angle view from Mars Express shows the complex structure of the Aureum Chaos region, and highlights the narrow elevated causeways left behind where most of the surrounding terrain collapsed.

Candor Chasma
6.6°S, 70.9°W

One of several individual canyons running through the broadest part of the Valles Marineris, Candor Chasma has been studied in detail by a variety of satellites and was once considered as a potential landing site for NASA's Curiosity lander. Its steep-sided walls are shaped by forms of erosion that do not require the presence of water, and the canyon rocks themselves are probably volcanic in origin, but sedimentary deposits in some parts of western Candor Chasma contain "hydrated" minerals such as sulfates that could only have formed in the presence of water. These so-called interior layered deposits (ILDs) may therefore record a time when large parts of the Valles Marineris were underwater. Nearby, researchers think they may even have identified rocks chemically transformed by the hydrothermal activity of hot springs.

Elsewhere in Candor Chasma, huge blocks of rubble mark the sites where ancient landslides brought rocks crashing down from the canyon rim, and delicate dune patterns show how prevailing winds sculpt the surface sand and dust. While Candor may have ultimately been overlooked in the selection process for Curiosity's landing site, it seems almost certain that it will be targeted by a future robot mission.

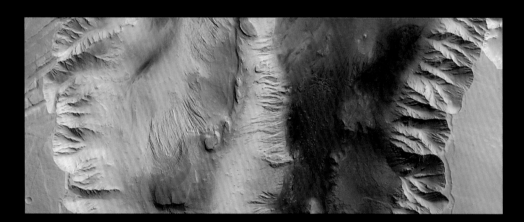

← An elevation map compiled from Mars Express observations reveals height variations around Candor Chasma. The deepest areas of the canyon (colored blue) are more than 30,000 feet (9,000 meters) below the bordering surface of the Tharsis rise.

↓ This perspective view, also generated from Mars Express data, looks across the canyon, roughly toward the northwest, revealing the complex structure of Candor Chasma's north wall.

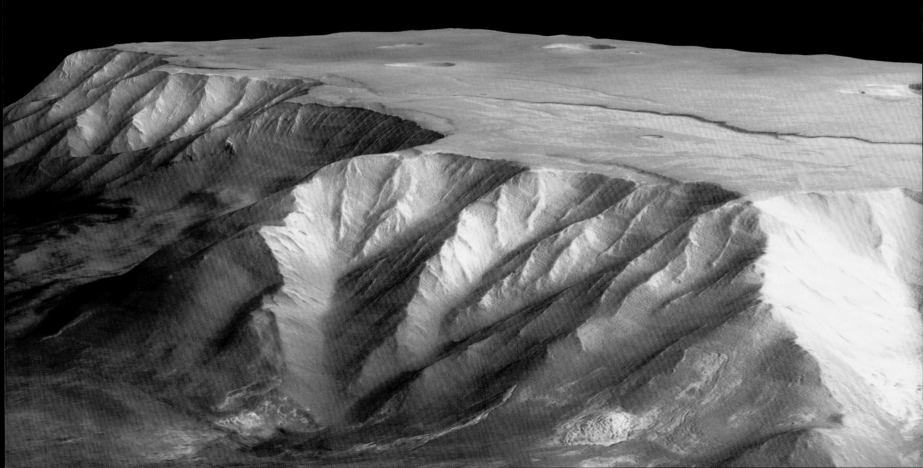

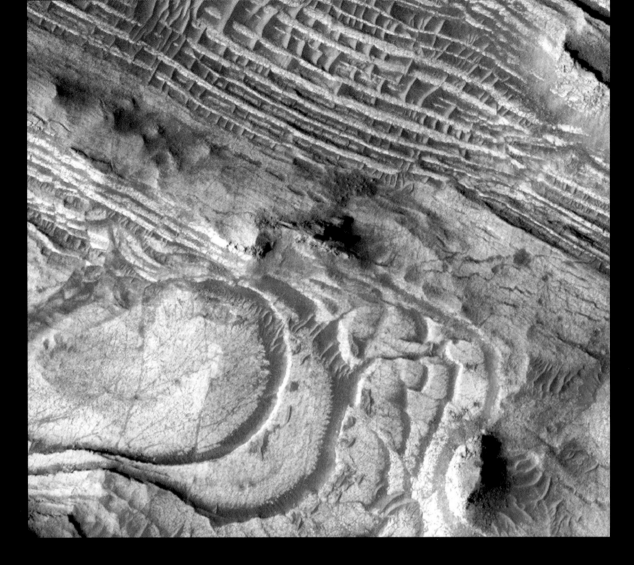

← A Mars Reconnaissance Orbiter close-up focuses on the tortured floor of western Candor Chasma, revealing signs of complex folding and faulting processes in the canyon's ancient history. Some of the layers may even echo patterns in the original sediments that were compressed to form the rock, such as dunes or wavelike ripples.

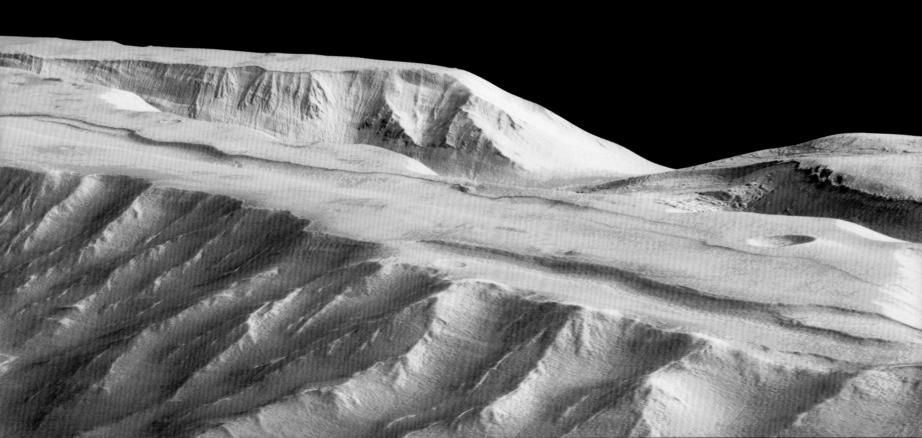

Meridiani Planum
0.2°S, 2.5°W

Situated at the western end of Sinus Meridiani (an albedo feature first identified in the 1870s by French astronomer Camille Flammarion), Meridiani Planum was chosen as landing site for the Mars Exploration Rover Opportunity following the discovery that its rocks contain a substantial amount of the mineral hematite. This iron oxide mineral is believed to only form in the presence of water, so its presence here was taken as strong evidence that the plain had a wet past—in fact, Meridiani Planum's hematite deposits are only rivaled by those seen in the Valles Marineris and Aram Chaos regions (see pages 84 and 88), neither of which provided a practical landing site for the delicate rover.

As its name suggests, Meridiani Planum is a smooth, gently rolling plain with a sandy soil and few large rocks (something that greatly assisted the identification of "Heat Shield Rock," the first meteorite ever found on another planet—see page 195). Instead, its surface is littered with small gray spherules nicknamed "blueberries," which turned out to be composed of pure hematite. These and other minerals in the nearby rocks suggest that the plain spent a long time underwater, allowing the hematite blueberries to form within sedimentary layers on a lakebed. Subsequently, the lake evaporated away and the surface sediment eroded, leaving the more durable blueberries scattered across the dry plain.

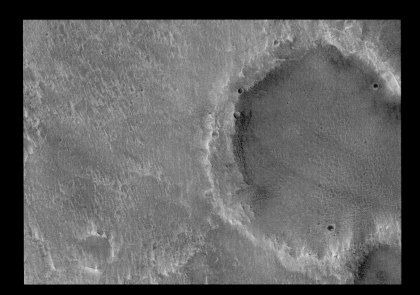

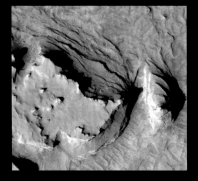

↑ Parallel wind streaks from small craters in northern Meridiani Planum are created by a prevailing wind that erodes bright material from the crater rims and deposits it downwind.

→ This exposed ridge in the nearby Terra Meridiani region is probably composed of sedimentary rocks laid down by a shifting former streambed. The relief of the region has now been inverted by erosion of the surrounding softer terrain.

↑ So-called "etched" terrain exposed in northern Meridiani Planum is believed to underlie the entire plain, and formed the base upon which later sediments (such as those around Opportunity's landing site) accumulated and eroded.

→ The bright flows and puzzling near-circular structures in this rock lie around 190 miles (300 kilometers) from Opportunity's landing spot, toward the west of Meridiani Planum. The bright material appears to be a relatively thin layer, and one theory suggests that the dark spots represent bubbles of fluid material that welled up from beneath it shortly after it was laid down.

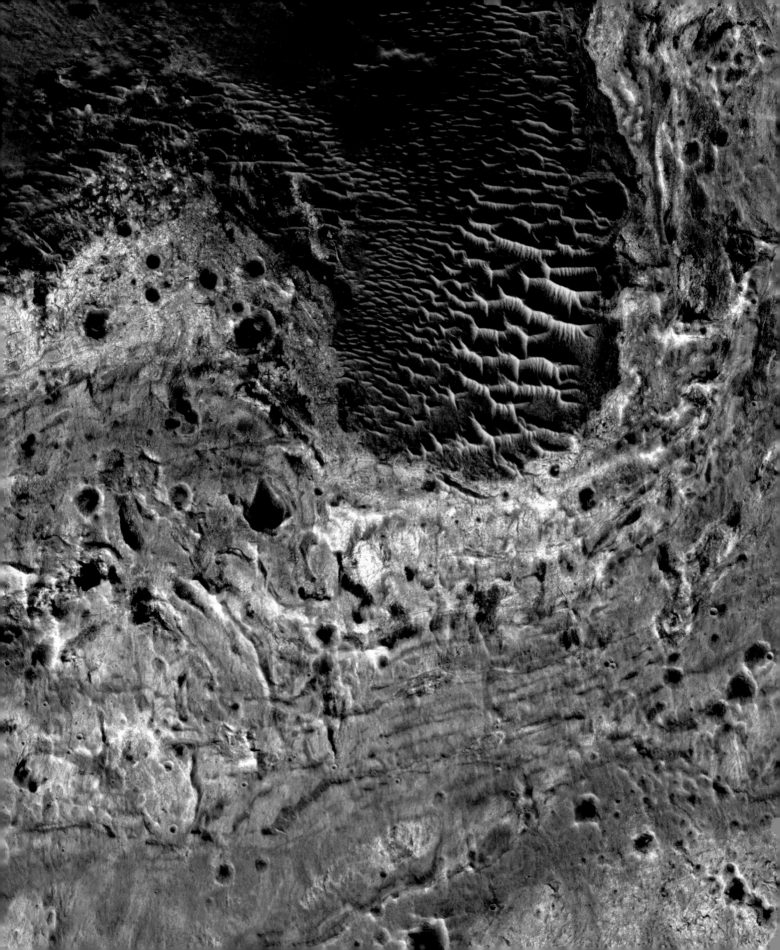

Arsia Mons
8.4°S, 120.1°W

The southernmost of the Tharsis Montes volcanoes, Arsia Mons is also the largest of the three, dwarfed only by Olympus Mons. Like its neighbors, it is a shallow shield volcano, rising to 5.5 miles (9 kilometers) above the surrounding Tharsis rise (12.5 miles or 20 kilometers above the Martian surface datum) across its 275-mile (440-kilometer) diameter. Arsia is the only one of the Tharsis volcanoes in the southern hemisphere, and it produces its own local weather pattern at the start of each southern winter—a spiral dust cloud that can tower up to 19 miles (30 kilometers) above the volcano.

The central caldera formed when the volcano's exhausted magma reservoir drained away about 150 million years ago and, with a diameter of 70 miles (110 kilometers), it is large enough to hold a series of smaller volcanic shields within it. The northwestern flank of the volcano, meanwhile, looks distinctly different from the other slopes, with a rougher texture that may have been carved by the action of glaciers. Elsewhere on the flanks, the roofs of now-empty lava tubes have collapsed to produce deep pits that were at first thought to be cave entrances.

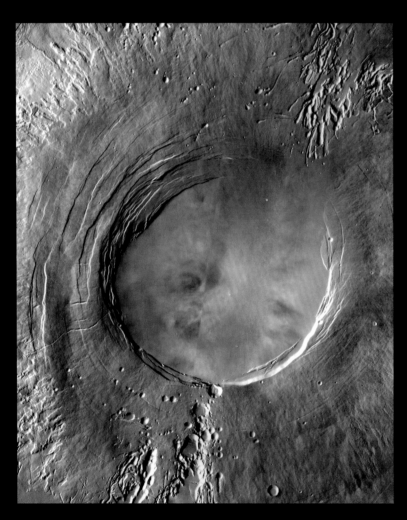

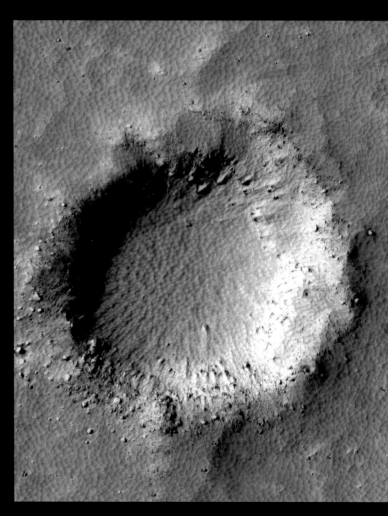

↑ This view of Arsia Mons combines several infrared images from the THEMIS instrument on NASA's 2001 Mars Odyssey probe. Patterns of fractures aligned with the neighboring Tharsis Mons volcanoes add weight to the idea that the three huge shields are a chain formed by a moving hot spot in the

↑ Mars Reconnaissance Orbiter's close-up view of a much smaller impact crater inside the enormous volcanic caldera emphasizes the amount of dust that accumulates in this elevated region of Mars, rapidly blanketing and smoothing out all but the starkest of surface features.

→ A pit on the western flank of Arsia Mons offers a rare insight into the structure of a volcanic shield, revealing dozens of layers of volcanic rock laid down in separate eruptions as the volcano rose to its present height. Frustratingly, as elsewhere on Tharsis, much of the detail is blanketed in dust.

Gorgonum Chaos
37.5°S, 170.9°W

A complex network of valleys in the Phaethontis quadrangle of the southern hemisphere, Gorgonum Chaos is best known as the site where the first evidence for Martian gullies was identified. Images of the region gathered by Mars Global Surveyor in 2000 showed alcovelike features emerging from just below the top of a valley, converging into deeper channels and then discharging in an area known as an apron on the valley floor (see page 59). While the gullies immediately suggested the behavior of liquid seeping out of an aquifer layer just beneath the surface, a variety of other mechanisms were proposed that could have produced them without the need for liquid water. More recent discoveries seem to have confirmed that a watery origin is most likely (see page 60).

Chaos features on Mars typically consist of flat-topped mesas separated by collapsed terrain in the form of shallow, flat-bottomed valleys and hummocky hills, as seen in both Aram and Aureum Chaos (see pages 88 and 140), and are usually assumed to have formed from catastrophic subsidence as water underlying the landscape escaped through an outflow channel. A different sort of mechanism seems to have been at work in the formation of Gorgonum Chaos, however, since many of its valleys are V-shaped, and there is no nearby outflow channel. Instead, some researchers think the entire region may be a result of wind erosion at work on a landscape built up from fine ash blown here from the Tharsis volcanoes.

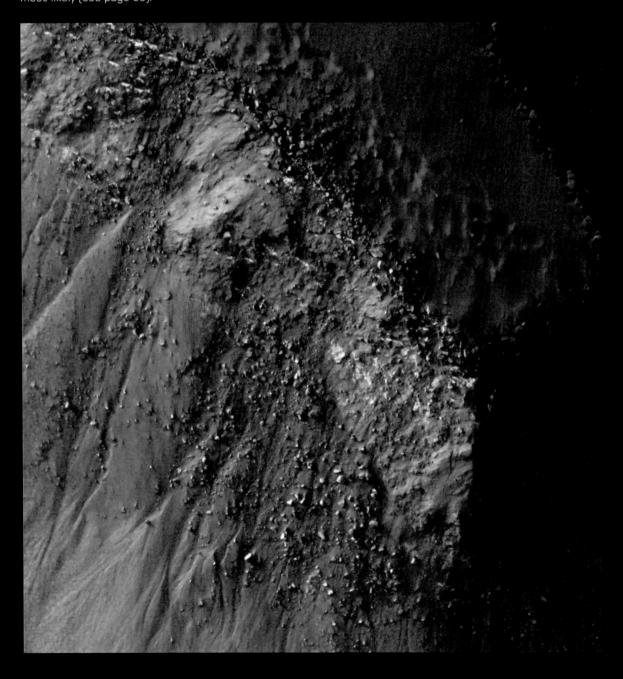

→ This famous image from Mars Global Surveyor provided the first evidence of gullies linked to the ongoing action of liquid water on the Martian surface. Dozens of distinctive fan shapes emerge from just beneath the surrounding surface layers along the south-facing slopes of this wind-sculpted valley.
← A Mars Reconnaissance Orbiter image focuses on the edge of a single mesa within the Gorgonum Chaos region, revealing the gully features in far more detail, and with greatly enhanced color differences. Again, the gully alcoves all clearly seem to originate from a well-defined layer just beneath the mesa surfaces.

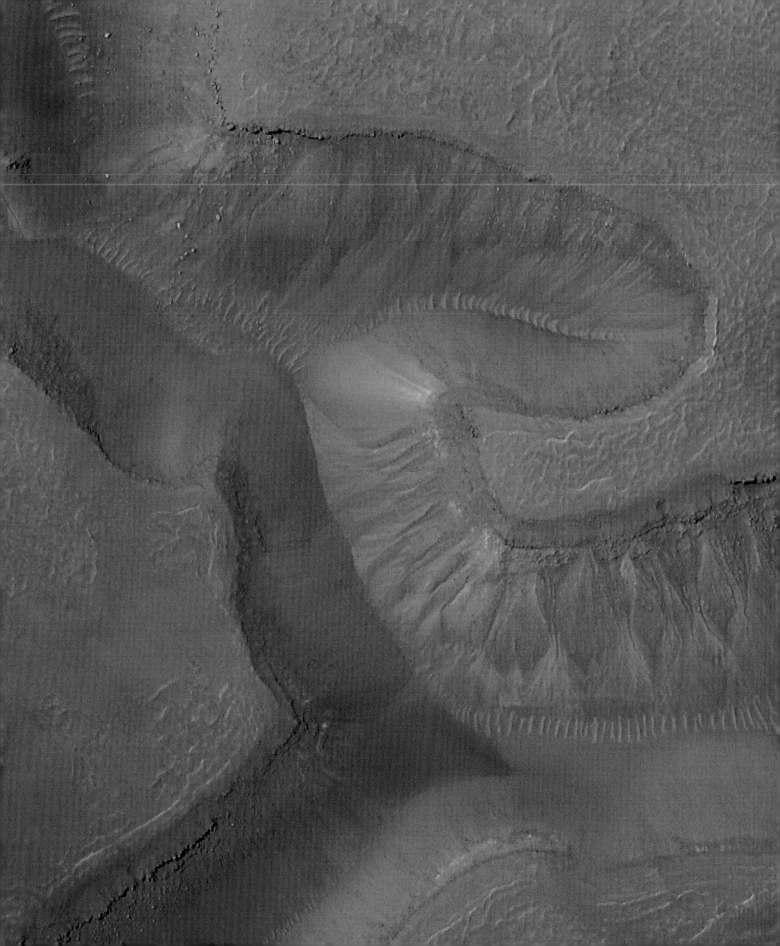

The most prominent feature on the entire Martian surface, Syrtis Major was identified as a dark triangle to the north of the equator by telescopic observers as early as the 17th century, but owes its current name to Italian astronomer Giovanni Schiaparelli (see page 14). Until quite recently, it was assumed to be a flat plain at a higher elevation than nearby Isidis Planitia (which lies directly to its west). However, the extensive topographic mapping carried out by Mars Global Surveyor revealed the truth —that Syrtis Major is in fact a low shield volcano, roughly 940 x 620 miles (1,500 x 1,000 kilometers) across. The volcano rises gradually to an altitude of 3.7 miles (6 kilometers) at its peak (with an average slope of just 1 degree), where a central depression containing the volcanic calderas, Nili and Meroe Paterae, sinks to 1.2 miles (2 kilometers) below its surroundings.

Moderate cratering across Syrtis Major suggests that the volcano was formed in the Hesperian period more than 2 billion years ago, as a result of eruptions that poured mile-thick streams of lava onto the Martian surface. The dark rocks formed by these basaltic lavas give the region its distinctive color, while the region's altitude apparently protects it from being overwhelmed with the ubiquitous red dust. Nevertheless, dust does accumulate and clear away from Syrtis Major in an apparently seasonal cycle, producing changes in the intensity of its darkness that are easily seen from Earth.

→ This Viking 1 mosaic shows the mottled surface of Syrtis Major in detail. The region owes its dark color to exposed volcanic rock with little overlying dust—the many streaks that fleck its surface are created by bright material eroded from crater rims and other raised features by prevailing winds.

↓ A topographic map compiled using the MOLA laser altimeter aboard Mars Global Surveyor reveals the shallow slopes of the Syrtis Major shield volcano, with the smooth basin of Isidis Planitia to the east. Despite this discovery, Syrtis Major is still currently named as a plain rather than a Martian volcano.

↓ Seen toward the right of the Martian disk in this Hubble Space Telescope image, Syrtis Major is the planet's most prominent surface feature after the bright polar caps. It was first identified in 1659 by Dutch astronomer Christiaan Huygens, who used its position to estimate the length of the Martian day.

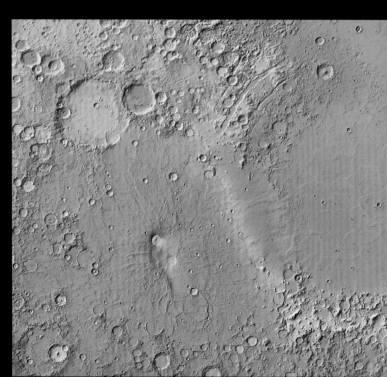

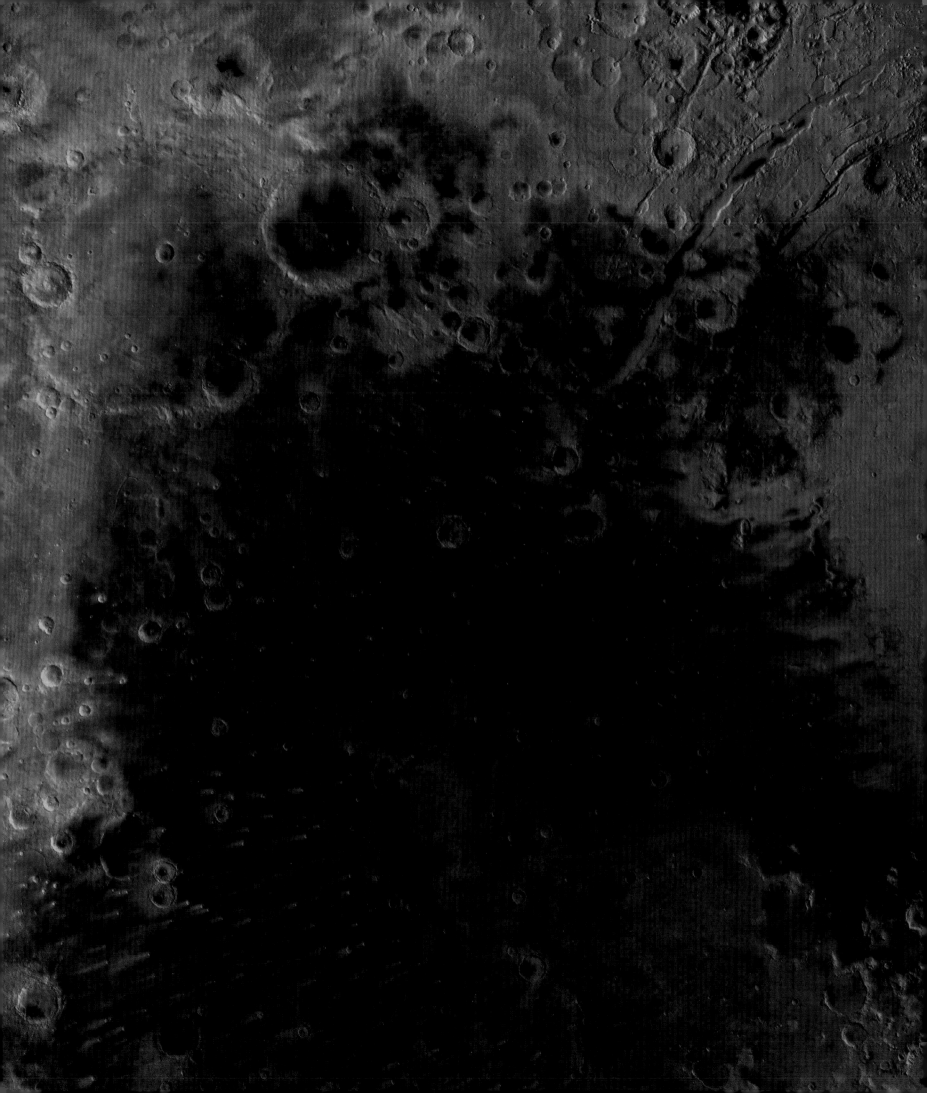

been an enormous depression.

Since then, layers of sedimentary rock have also accumulated within the basin, creating today's plain. Situated at mid-latitudes, the area is cold enough to be covered with frost for much of the Martian year, and permafrost persists beneath the surface. Continued expansion and contraction of the ice

crusty surface, with loose soil apparently removed from around them. Analysis of these rocks showed that they were largely volcanic in origin and probably originated as ejecta flung out during formation of a nearby 65-mile (105-kilometer) impact crater called Mie.

← This Mars Reconnaissance Orbiter (MRO) view shows the unusual large-scale surface features of Utopia Planitia, thought to be formed by the uneven sublimation of ice from within the soil. Researchers call this kind of landscape a "thermokarst."

→ A close-up from MRO's HiRISE camera focuses on small-scale polygons formed by seasonal expansion and contraction of the permafrost. The polygons are typically around 33 feet (10 meters) across.

↓ This image from the Viking 2 lander was the first color view of Utopia Planitia's surface. The largest rock to the left of center is about 40 inches (1 meter) across, and a shallow trough, perhaps one of the polygonal cracks seen in orbiter images, runs diagonally from upper left to lower right beneath it.

This large highland impact basin is second only in the southern hemisphere to the Hellas basin, with a total diameter of around 1,100 miles (1,800 kilometers) and perhaps even more. The central plain is roughly 400 miles (650 kilometers) across, and 3.2 miles (5.2 kilometers) below its surroundings, but it is encircled by several concentric mountain ranges and massifs.

Argyre is thought to have formed during the early Noachian period of Martian history, perhaps around 4 billion years ago, and subsequently acted as a drainage basin for water flowing out of the highlands. The water may have begun life as natural precipitation during the warmer and wetter Hesperian period, or could have been released through the melting of an extensive south polar ice cap. Wherever it came from, the water seems to have accumulated through three major channels run into Argyre from the south and east, before draining out along a fourth that flows out to the north, toward the lowland area of Chryse (see page 126) and the hypothetical northern sea beyond.

The basin features many water-related features ancient and modern, ranging from sedimentary lake deposits through glacial features to gullies, perhaps produced by recent liquid water flow.

↓ This astonishing Mars Express view shows the northwest corner of Argyre Planitia at left, with a large crater called Hooke, 86 miles (138 kilometers) in diameter, to the right. This corner of the vast Argyre plain gleams in the sunlight thanks to a thin coating of carbon dioxide frost.

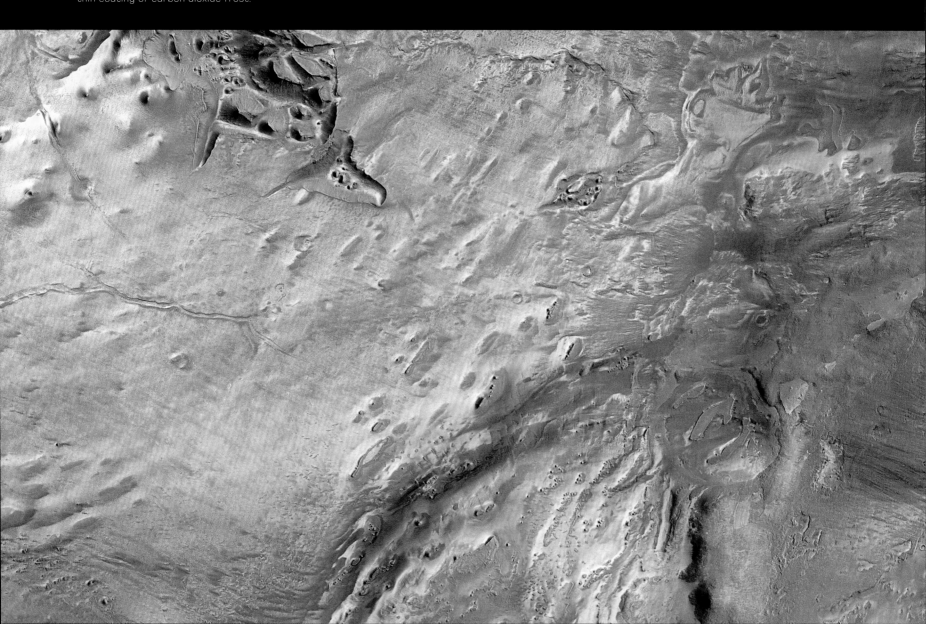

→ This spectacular Viking view looks east across the Martian highlands and reveals a rare profile of the thin Martian atmosphere. Argyre Planitia lies in the foreground, with the distinctive "Happy Face" crater Galle at center left.

This broad channel runs for 1,050 miles (1,700 kilometers) out of the southern highlands with an average width of 15.5 miles (25 kilometers), broadening into a much wider deltalike region as it descends into Chryse Planitia. Discovered by the Viking orbiters in the 1970s, it was one of the first major Martian outflow channels to be recognized and was the landing site for the Mars Pathfinder mission in 1997.

A series of distinctively streamlined islands in the valley mouth suggest that Ares Vallis was created as a result of sudden and catastrophic floods rather than a long-term steady flow of water. The valley originates close to a broad sunken region known as Iani Chaos, and this is likely to have been the source of the floodwaters—like other chaos regions, Iani probably formed through subsidence after huge amounts of subsurface ice or water were drained away. Other proposed sources for the torrents that carved out Ares Vallis include one of a group of large equatorial lakes that may have once ringed the Martian equator, or the much larger and more distant reservoir of Argyre Planitia (see page 154).

→ This perspective view of Ares Vallis from Mars Express highlights streamlined islands in the broader valley, indicating the direction of the floods that carved out this broad Martian valley.

← A topographic map (far left) and an orbital view from Mars Express show details around the 20-mile (32-kilometer) crater Oraibi. The south wall of this ancient crater was entirely washed away during catastrophic floods, swamping the crater interior with sediments that now form its shallow, flat floor.

↓ This panoramic view from the Mars Pathfinder base station (see page 176) reveals a landscape scattered with small rocks. Ares Vallis was selected as Pathfinder's landing site partly because of the variety of rocks carried here from the nearby highlands.

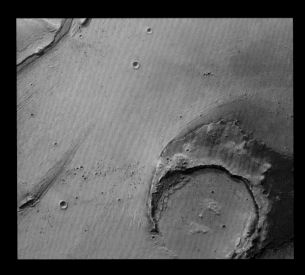

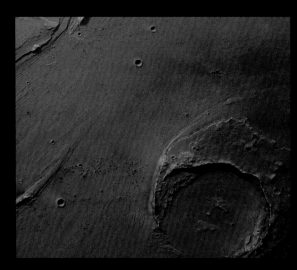

Capri Mensa
14.0°S, 47.4°W

At the eastern end of the Valles Marineris complex, a long, straight canyon known as Coprates Chasma opens out into the broader and more chaotic region known as Capri Chasma. Here, large parts of the canyon walls collapsed in the distant past to leave isolated steep-walled mesas such as Capri Mensa standing several miles above the canyon floor.

The geography of Capri Chasma seems to have encouraged the buildup of sedimentary rock outcrops known as interior layered deposits (ILDs). In places, they are probably gradual accumulations of volcanic ash, but in others they show distinct signs of periodic layering.

This suggests that they may have formed within standing bodies of water that accumulated and then disappeared repeatedly in a cycle linked to Martian climate variations.

Satellite mapping also shows that many of the layered deposits in this area have intriguing mineral compositions, including sulfate minerals and even hematite (see page 144) that could only have been created in the presence of water. The hematite may consist of "blueberries" similar to those found by the Opportunity rover on Meridiani Planum (see page 192), suggesting similar processes at work in both locations.

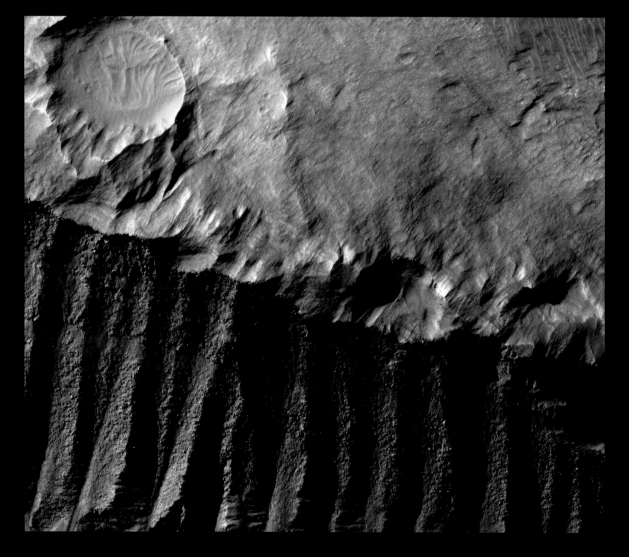

→ A striking image from Mars Reconnaissance Orbiter's HiRISE camera reveals details along the wall of a large impact crater in Capri Mensa. A thin layer of smooth material on the crater walls has worn away in places to reveal rocky material beneath, while light-toned rocks with a probably sedimentary origin are exposed amid drifts of dust on the crater floor.

← This HiRISE view focuses on the edge of an interior layered deposit (ILD) close to Capri Mensa. Stratified sediments can be clearly seen at the bottom of the picture where the side of the ILD has suffered a later landslip.

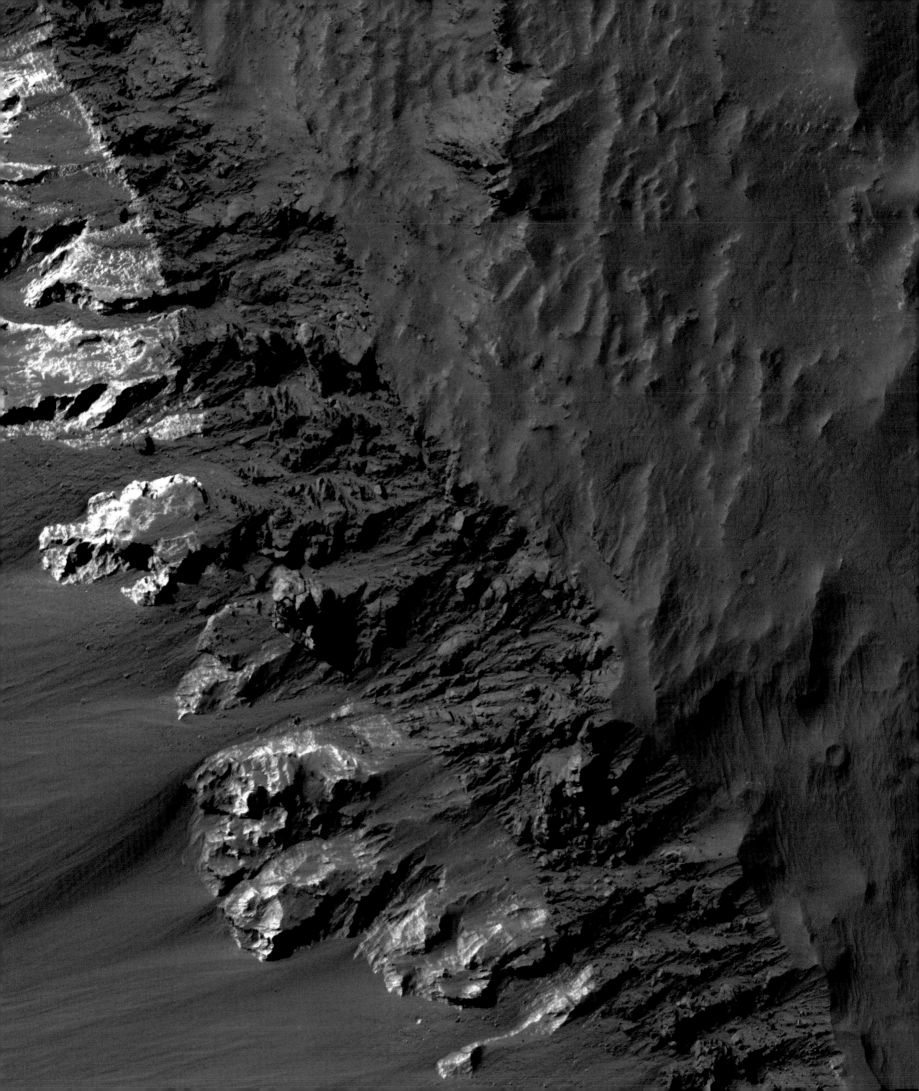

This long, broad canyon runs parallel to the northwestern edge of the Isidis basin, a circular depression in the Martian crust some 930 miles (1,500 kilometers) across that formed when a large asteroid slammed into the planet about 4 billion years ago. Hundreds of miles long and more than 15.5 miles (25 kilometers) wide, the canyon is a fault created as the surrounding terrain cracked and subsided.

Since its formation, Nili Fossae has had a long and complex geological history. Its floor was flooded by lava believed to have flowed from volcanic eruptions in the Syrtis Major region (see page 150), and subsequently blanketed by clay-rich "ejecta" —material that was probably thrown out during formation of the nearby 40-mile (65-kilometer) Hargraves crater. Since then, the ubiquitous

Martian dust has accumulated in the canyon where it is blown into often-spectacular dune fields.

The detection of suspected clays in this area (minerals that could only have formed in the presence of water), as well as the exposure of some of the earliest Martian rocks in the sides of the fault, made Nili Fossae a potential target landing site for NASA's Curiosity rover, although it did not make the final shortlist. More recently, however, Nili Fossae has become a focus for interest in potential Martian life, since it appeared to be the origin of a large plume of methane gas reported from Mars Express observations in 2009 (see page 150).

↓ This Mars Reconnaissance Orbiter image peers into the depths of the Nili Fossae, revealing basaltic and clay-rich materials. The volcanic rock and dust appears blue in this enhanced-color image, while clays, almost certainly formed when the area was submerged beneath water, have lighter orange and yellow tones. The Mars Express overview of the region shows the broader region in more realistic colors.

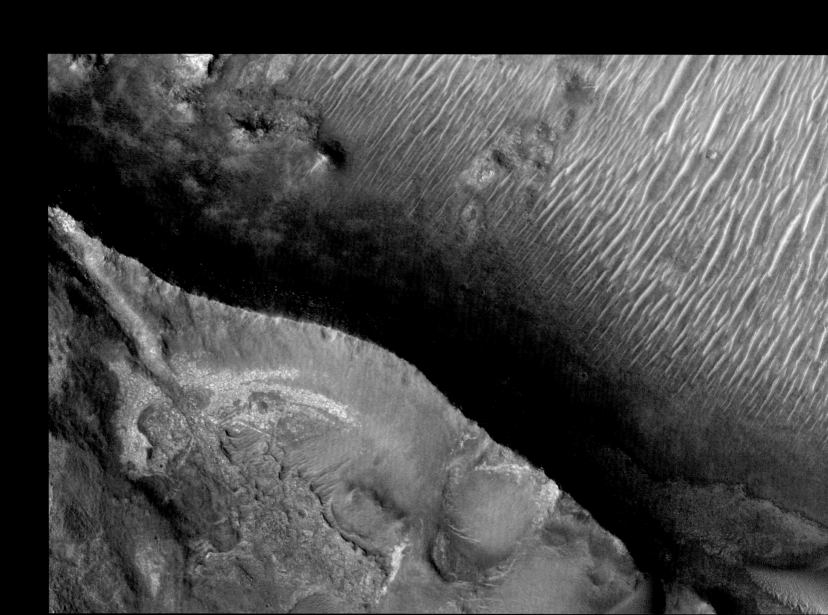

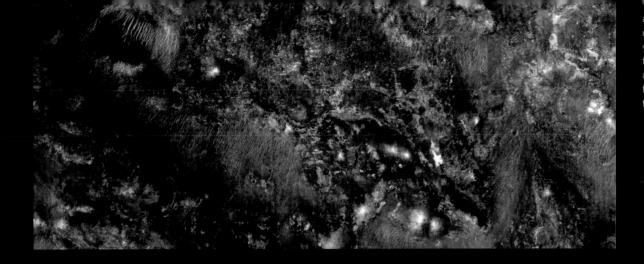

The jumbled rocks in this image are ejecta from a meteorite impact close to Nili Fossae many millions of years ago. Enhanced colors indicate the broad variety of different rocks and mineral compositions to be found among materials ejected from the crater.

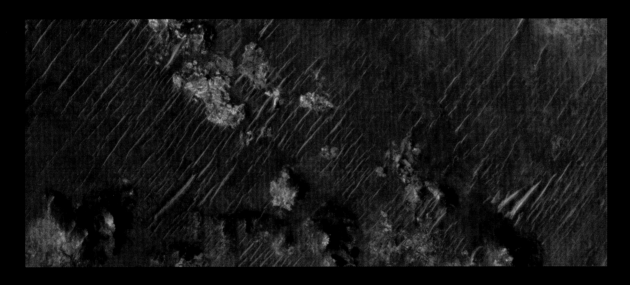

← A highly color-enhanced view of the Nili Fossae shows basaltic volcanic rocks in purple, clays in orange and unaltered pyroxene, one of the unaltered minerals of the underlying Martian bedrock, in green.

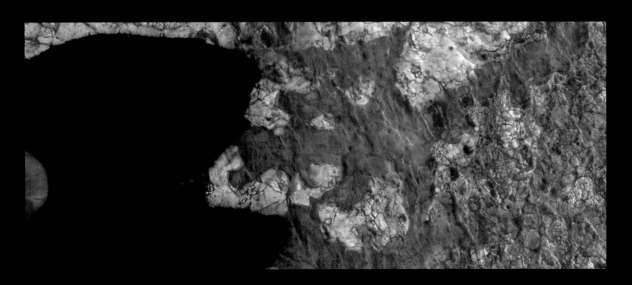

← A pair of distinctively shaped barchan dunes sweep over the lighter surface of a lava flow in Nili Patera, a volcanic caldera close to Nili Fossae. The dark sand is probably also volcanic in nature, but its origins are uncertain.

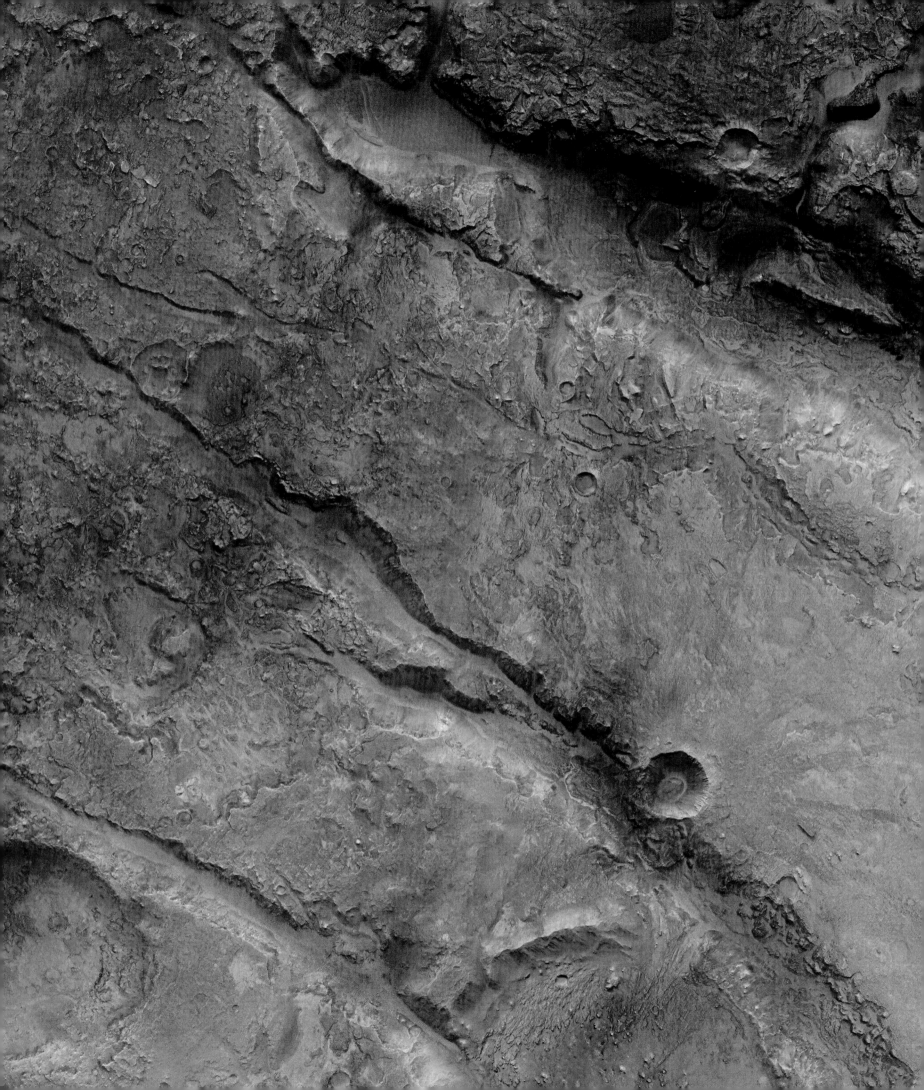

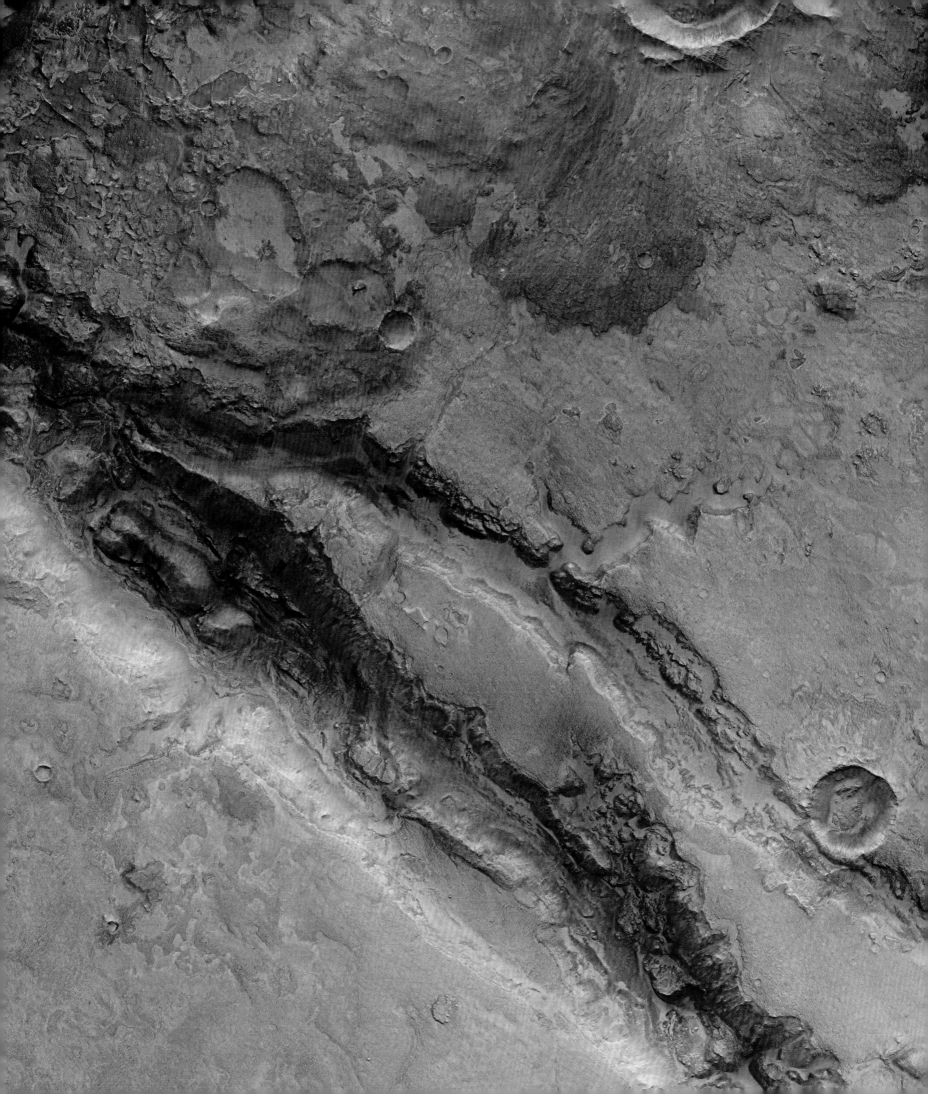

in the Martian southern hemisphere, Reull Vallis is a steep-sided, winding canyon. Its features suggest that it was carved by a river that ran for a considerable period of time, rather than a brief, catastrophic flood. The river would have drained water from a broad area of the highlands toward the deep floor of the Hellas basin at its western end. Numerous tributaries join the main canyon system at various points along its length, and the canyon's winding path is also unlikely to have been eroded by sudden floods. Based on the requirement for surface-flowing water and counts of craters

perhaps 2.5 billion years ago.

However, one of the valley's most striking features shows that it has been modified more recently: patterns of parallel grooves running along the floor in places were probably scoured by glaciers that took the same path. Evidence of glaciers is widespread across these mid-latitude highlands, with similar features found in some nearby craters, and glaciers themselves, in the form of so-called "lobate debris aprons" (see page 122), hidden just below the surface in many places.

↓ An elevation map of Reull Vallis, derived from Mars Express data, reveals the contrast between the steep-sided valley and the rounded, mountainous and cratered highlands to the west. Narrow channels within the main valley are a key piece of evidence that it was carved by a long-lived river rather than a brief flood.

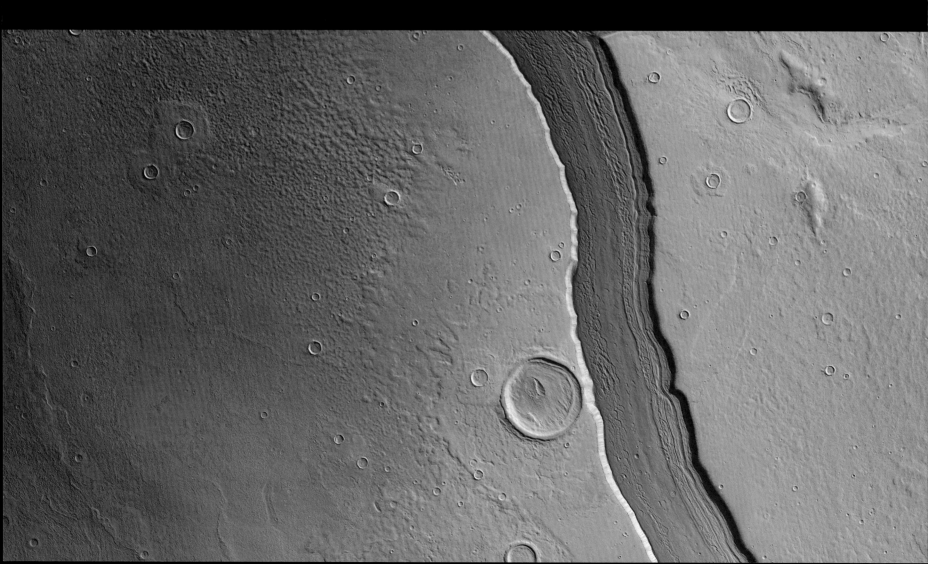

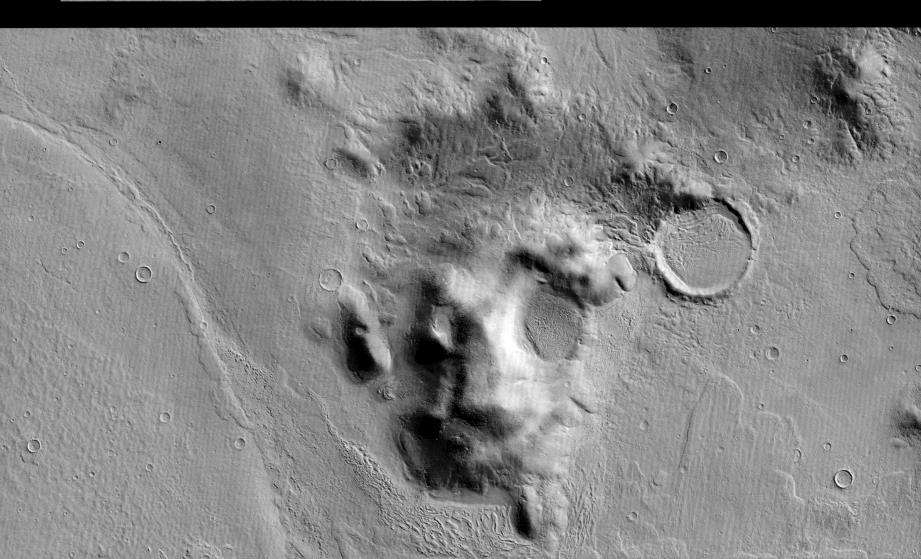

4 Explorers on the Red Planet

While human adventurers have yet to set foot on Martian soil, we have sent our ambassadors ahead, in the form of robot space probes. These orbiting satellites, surface laboratories, and intrepid rovers have sent back a wealth of information to Earth, transforming our understanding of Mars forever and paving the way for that inevitable day when humans make the journey for themselves.

→ A view across the rubble-strewn Ares Vallis around Mars Pathfinder's landing site.

Early probes

The fascination exerted by the Red Planet made Mars an obvious target for exploration from the very dawn of the Space Age in the late 1950s, but early Martian space probes frequently met with frustration and failure, and it was not until the early 1970s that the first probe finally entered orbit.

Mariner 4

Mariner 6

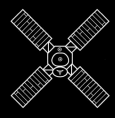

Mariner 9

The early stages of the Space Race between the United States and the Soviet Union were largely governed by the nature of the rival nations' rocket technology, developed largely to deliver nuclear missiles and other weapons across intercontinental distances. More compact US warhead technology left them with relatively small missiles, so that early space exploration efforts were limited in their ability to launch manned capsules into orbit or larger space probes further into the solar system. The Soviets' cruder and heavier warheads required larger missiles, so from the outset they were equipped with huge Molniya rockets capable of launching robot probes to the Moon and nearby planets.

So it was that the Soviets made the first attempt to reach Mars, beginning what would be a long catalog of failures and disappointments. The two Mars 1M probes, launched in 1960, both failed to even achieve Earth orbit due to a malfunction in the upper stage that would eventually have boosted them on their way to Mars. Two years later, at the next opposition (close approach between Earth and Mars) they tried again, only to meet with more frustration: the Mars 1 probe was successfully

launched onto a trajectory that flew past the planet at a distance of 120,000 miles (193,000 kilometers), but all communication was lost with the probe around 66.5 million miles (107 million kilometers) from Earth and still some way from Mars. A further two attempts at launching Mars probes that year (another flyby mission and a lander) both failed shortly after launch.

By 1964, US rocket technology had developed enough for NASA to attempt its own Mars mission. The US space agency's strategy involved launching pairs of identical space probes in each biennial launch window, and it paid off immediately— while the first US Mars probe, Mariner 3, failed during launch in early November, its twin sibling, Mariner 4, launched successfully three weeks later and maintained contact with Earth through to its Martian flyby on July 14, 1965. Mariner 4 successfully sent back the first close-up images of another planet, but its flight path over the southern highlands inadvertently led to a deceptive picture of a heavily cratered, Moonlike world that seemed to have changed little through its history. The probe also returned measurements of the Martian surface temperature and magnetic field,

↑ Mariner 4's octagonal, drum-shaped body was just 4 foot 2 inches (1.27 meters) across, with solar panels extended in an X-shape around it. Unusually, solar pressure vanes were attached at the end of each "wing" to assist steering using the pressure of particles from the Sun.

→ This fuzzy image from Mariner 4 captures the outline of a 94-mile (151-kilometer) crater in the southern highlands, subsequently named the Mariner crater. During its flyby, the probe captured 21 complete images covering roughly 1 percent of the Martian surface.

approaching to within 6,115 miles (9,840 kilometers) of its target. Meanwhile, the Soviets lost a further two probes in the 1964 launch window.

Two more NASA probes (Mariners 6 and 7) successfully flew past Mars and returned images during the 1969 launch window (again passing coincidentally over the cratered southern highlands), and it was clear that the next step would be to put a probe in orbit around the planet during the 1971 launch window. Here the strategy of building probes in pairs paid off once again, as Mariner 8 failed during launch, but Mariner 9 successfully lifted off from Cape Canaveral on May 30, 1971 and entered Martian orbit on November 14 —the first spacecraft ever to orbit another planet.

Although crude by later standards, Mariner 9's suite of scientific instruments transformed our view of Mars. These included a hugely improved camera for observing the surface—though still only at

relatively large scales—and infrared and ultraviolet spectrometers for analyzing the composition of surface minerals. Famously, the probe arrived during a major dust storm (still the largest and most persistent ever recorded) and the scientific mission was delayed for a couple of months. As the dust began to settle, the first features to emerge from the haze were the peaks of Olympus Mons and the Tharsis volcanoes.

Throughout 1972, Mariner 9 sent back the first images of far more complex terrain away from the cratered highlands, including ancient Martian riverbeds, the enormous canyon system now known as the Valles Marineris, and even clouds in the Martian atmosphere. Although the picture returned of a dry, cold Mars was a disappointment to some, the mission itself was a resounding scientific and technological success that would inspire an even more ambitious follow-up.

↓ ↓ At the time, Mariner 7 captured the sharpest images of the entire Martian disk during its August 1969 flyby. The polar caps stand out prominently, while the circular feature above center was named Nix Olympica, subsequently identified as the enormous volcano Olympus Mons.

↓ Mariner 9 was larger and heavier than previous flyby probes, with most of its extra weight due to the engines needed to enter a controlled Martian orbit. A "scan platform" beneath the drum-shaped body, rotating on two independent axes, enabled instruments to target specific areas of the surface.

↓ ↓ A complex branching valley, named Nirgal Vallis after the ancient Babylonian name for the planet, taken by Mariner 9. One of the longest valleys on Mars, Nirgal's short tributaries suggest that its waters may have originated from ice or liquid beneath the surface.

↓ Mariner 9's most famous discovery was the Valles Marineris canyon system that bears its name. This image, taken early in the mission, captured the enormous scar in the Martian crust as it emerged from the enormous dust storm that had cloaked the planet since Mariner's arrival.

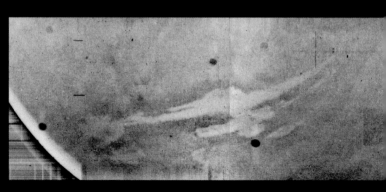

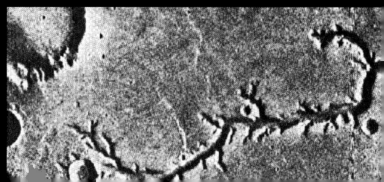

From the first Martian explorers to the advanced rovers of today—this astonishing image from the Curiosity rover (see page 202) reveals sedimentary rocks around the rover's landing site in Gale crater. An outcrop known as Yellowknife Bay is being exposed by steady wind erosion—foreground rocks are about 50 feet (15 meters) from Curiosity's Mast Camera. As shown here, the terrain of Mars is far from smooth, and this is just one of the challenges that has faced scientists during the development of the Mars rovers.

Viking orbiters

Even while NASA's first wave of Mariner and Pioneer space probes were venturing out to make the first reconnaissance of our solar system, plans were afoot for a more ambitious second wave of exploration.

Viking 1

With the Apollo Moon program attracting huge interest and investment, but the first lunar landing still some way off, a manned moonbase and an expedition to Mars seemed like an inevitable next step. Any Mars expedition would require extensive foreknowledge of surface conditions, and so development on the Voyager Mars Program began in 1966. Forming part of the wider Apollo Applications Program, the planned series of missions would take advantage of Apollo-era technology such as the enormous Saturn V rocket and the sophisticated Surveyor robot lander (modified for descent into the thin Martian atmosphere). Once again, NASA would launch missions in pairs—this time using a single Saturn rocket to send two separate spacecraft on their way to Mars. Each spacecraft, with a mass of up to 12 tons, would separate on arrival into an orbiter that would continue the task of surveying and mapping the Red Planet from space, and a lander that would descend to the surface and report on the conditions it found, carrying out studies of the soil and atmosphere.

By 1967, plans were underway to launch a pair of probes in each of the 1973, 1975, 1977, and 1979 launch windows, with the landers carrying ever more sophisticated experiments at each launch. But then the first of many NASA budget cuts saw the entire Apollo Applications Program axed, taking with it the Voyager missions (though the name was later recycled for NASA's 1970s and 1980s flyby missions to the gas giant planets) and the ambitious dreams for lunar bases and manned exploration of the deeper solar system.

Nevertheless, further unmanned exploration of Mars remained a subject of intense interest, and NASA's Jet Propulsion Laboratory soon found a cheaper way of fulfilling many of the same objectives. Voyager's successor, known as the Viking Program, would pair an orbiter spacecraft (modified from the Mariner 8/9 design) with a significantly lighter and simpler Mars lander. Once again, two missions would be launched at the same time to improve the chances of a success, but off-the-shelf Titan III-E/Centaur rockets would be used to blast them individually into space, instead of using the enormous and hugely expensive Saturn V. Despite these economies, the Viking Program had still ballooned in cost to more than a billion dollars by the time the spacecraft were launched in August and September 1975, making it the most expensive mission ever launched to Mars—a record that it still holds today after inflation is taken into account.

Fortunately, the Vikings delivered a huge scientific return on this enormous investment. Both spacecraft performed more or less flawlessly, with Viking 1 entering orbit around Mars on June 19, 1976, and Viking 2 following it on August 7. Each mission spent roughly a month carrying out an initial survey of potential landing sites before releasing its lander to parachute to the surface (see page 174). Once relieved of their burden, the two orbiters began the task of building up a comprehensive and detailed map of the entire planet. Each was equipped with a pair of "vidicon" cameras (essentially television cameras) for transmitting electronic images of the visible surface back to Earth, as well as an infrared spectrometer for mapping atmospheric water vapor and a radiometer for measuring surface temperatures. Viking Orbiter 2 operated until July 1978, when it was deliberately boosted to a high, safe orbit and turned off after developing a propellant leak. Viking Orbiter 1, meanwhile, continued to function for another two years before being put in a similar parking orbit.

The Viking imagery of the Martian surface was far more comprehensive and detailed than that collected by Mariner 9, and as the orbiters moved beyond their initially planned operations, NASA engineers dropped them into orbits that brought them considerably closer to the Martian surface, allowing them to collect pictures with roughly four times Mariner's resolution (down to 82 feet or 25 meters per pixel of image data). Stitched together by computer, the Viking dataset formed the basis of a global map that is still in use today. It revealed a wealth of new features, including flood channels, rivers, and drainage systems that bore an obvious resemblance to similar features on Earth.

→ One of the Viking program's greatest achievements was imaging almost the entire Martian surface in color and at high resolution. Hundreds of images were stitched together to form detailed mosaics of entire planetary hemispheres, such as this one, representing the view from above the Valles Marineris at an altitude of 1,550 miles (2,500 kilometers).

← An artist's impression of one of the Viking orbiters releasing its aeroshell-clad lander close to apoapsis—the most distant point in its orbit around Mars, around 19,000 miles (30,000 kilometers) above the surface.

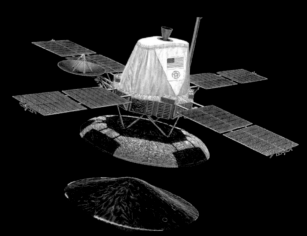

Viking landers

The Viking landers were the first spacecraft to land on the surface of Mars. Following a month of orbital reconnaissance, they were released from their parent orbiters and descended to the surface, targeting areas of the lowland plains on opposite sides of the planet.

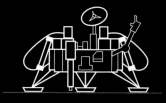

Viking Lander

Planning a controlled descent into the Martian atmosphere proved to be a challenge for the mission designers, just as it has been ever since. This was a major reason that the mass of each lander was reduced to a relatively lightweight 1,258 pounds (572 kg). The lander was protected on its underside by a heat shield, in the form of a shallow cone, to withstand the heat from entry into the Martian atmosphere and slow the vehicle to a point where a descent parachute could deploy at an altitude of around 20,000 feet (6,000 meters). Finally, since a parachute alone was a relatively inefficient brake in the thin Martian air, carefully mounted retrorockets on the lander fired at an altitude of 5,000 feet (1,500 meters), lowering the lander to the surface with a gentle bump. One significant challenge for the rocket designers was to place them in such a way as to leave the surface of the landing site relatively unaffected and suitable for study.

Viking Lander 1 touched down successfully in the Chryse Planitia region on July 20, 1976 and began sending back the first image of the Martian surface

almost immediately. The landing site was selected largely because it lay in the midst of the Chryse outflow region (see page 126)—an area that showed signs of catastrophic flooding in its past and which might therefore preserve evidence of Martian water. Estimates of the terrain roughness were made using Earth-based radar, but the exact landing target was changed at short notice after new orbital images revealed it to be rougher than expected.

Viking Lander 2, meanwhile, targeted the Utopia Planitia region of the northern plains—thought to be a relatively unaltered area of the original volcanic plain material. Once again the precise landing spot was changed at the last minute, but the spacecraft touched down safely on September 3, 1976. Despite this, the Utopia landing site turned out to be remarkably rocky on the smallest scales (see page 152) and would probably not have been selected if higher-resolution pictures had been available to show the dangers of the terrain.

↑ This Viking Lander 2 mosaic shows scattered rocks around the landing site in Utopia Planitia—a location that had been assumed to have smooth sandy dunes. Deep shadows beneath many of these stones suggest that wind has stripped away much of the underlying sand to leave them perched.

→ This unspectacular image of Viking Lander 1's foot was the first clear image ever sent back from the surface of another planet. The probe was programed to take and transmit an image just a few minutes after landing as a precaution, since earlier Soviet landing attempts had failed before they could achieve this.

← This Viking Lander 2 image, taken in May 1979, shows a thin layer of water-ice frost on the surface of Utopia Planitia at the onset of the mission's second Martian winter. The obvious slope of the image is due to Viking 2's uneven landing— by chance, the spacecraft touched down with one foot on a rock, tilted at an angle of about 8°.
↙ NASA subjected the Viking landers to intensive testing during the design process—this test model was photographed in the "Mars Yard" at NASA's Jet Propulsion Laboratory in California in 1975. Despite precautions, however, there were concerns that the Martian surface might be covered with deep layers of quicksand-like dust waiting to swallow up any vehicle that attempted a landing.

Although the landers were incapable of further movement after landing, they carried a sample arm that allowed them to scoop up nearby soil for analysis, as well as a complex suite of instruments whose development had swallowed much of the Viking Program's enormous budget. These included spectrometers for soil analysis, a simple magnetometer, seismometers for measuring potential Marsquakes, a weather station mounted on an extending boom, panoramic cameras for imaging the surface and, perhaps most importantly of all, an experiment for detecting possible signs of microbial life in the Martian soil. The landers were capable of communicating both with their respective orbiters far above using an S-band radio antenna and directly with Earth through a steerable parabolic antenna.

Despite the targeting of two seemingly distinct areas, the soil samples studied at both sites proved to be surprisingly similar, with large quantities of silica and iron suggesting that the soil had formed from the weathering of volcanic basalt rock. Other elements, such as magnesium, sulfur, aluminum, and calcium, were also common at both sites, but alkali metals were less common than on Earth. The general conclusion was that soil sampled on opposite sides of the planets came from essentially similar wind-blown dust particles.

The landers carried four distinct experiments to identify life: the Gas Chromatograph—Mass Spectrometer to evaporate material out of the soil and study its chemical composition; the Gas Exchange experiment to look for signs of organisms absorbing gas as part of their metabolic processes;

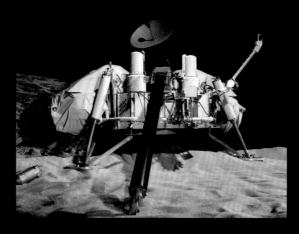

the Labeled Release experiment using radioactive carbon-14 to look for signs of nutrient absorption and release; and the Pyrolitic Release experiment looking for release of organic chemicals when soil was burned. To general surprise, no sign of any "organic" (carbon-based) chemicals was found— despite the fact that such chemicals are common in asteroids and comets across the solar system. Only the Labeled Release experiment produced hints of a possible organic reaction in the soil, but the result proved impossible to replicate and remains controversial even today (see page 73).

Both landers far outlasted their expected operational lifetime—Viking 2 lasted until April 1980 when its batteries failed, while Viking 1 was still operational in November 1982 when a faulty command from mission controllers resulted in a permanent loss of radio contact.

Following a string of delays and failures, NASA's Mars Pathfinder was to be a make-or-break mission, intended to put a small robotic rover on the Red Planet for the first time and pave the way for future, more ambitious, missions.

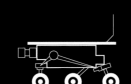

Pathfinder

Sojourner

Despite the huge success of the Viking missions, it was to be two decades before the next successful Mars landing. Throughout the late 1970s and 1980s, NASA's attention was largely directed closer to home, focusing on its ambitious Space Shuttle program while a series of Soviet missions met with limited success. Once the Shuttle was up and running, US interest in Mars returned, but the 1992 Mars Observer mission failed due to a loss of communications, and it was not until 1996 that the next successful missions were launched. These were the Mars Global Surveyor orbiter spacecraft (see page 178) and a compact but ambitious lander called Mars Pathfinder.

Pathfinder was designed to spearhead not only a return to Mars, but also a new philosophy of "faster, better, cheaper" spacecraft design following the failure of some hugely complex and expensive NASA probes, including the Mars Observer. Developed for US$150 million, it would provide a "proof of concept" for technologies that have shaped our exploration of the Red Planet ever since.

With a total mass of 605 pounds (275 kg), the lander was light enough to launch with a relatively inexpensive Delta II rocket on December 4, 1996. After an uneventful journey, it entered the Martian atmosphere on US Independence Day, July 4, 1997. Pathfinder was encased in a two-part "aeroshell" whose lower half functioned as a heat shield during its descent through the upper atmosphere. A parachute designed to cope with supersonic speeds was then released from the "backshell" to slow the descent still further, while the lander was lowered on a 66 foot (20-meter) cable. A little over

1,150 feet (350 meters) above the ground, a series of balloon-like airbags inflated to protect the lander, before a final burst of deceleration courtesy of retrorockets in the backshell that fired at an altitude of 321 feet (98 meters). The last stage of this complex descent was to release the lander at an altitude of 69 feet (21 meters), allowing it to fall the remaining distance and eventually bounce to a halt. Onboard sensors then deflated the airbags in sequence to ensure the lander settled in an upright position, and the lander then opened its three winglike solar panels to begin operations.

The main Pathfinder lander carried three main instruments—a stereoscopic camera mounted on a telescopic pole, the Atmospheric Structure Instrument (which recorded changing conditions during the lander's descent in order to build

↓ This photograph from the Mars Pathfinder base station shows the Sojourner rover shortly after it rolled onto the surface of Ares Vallis for the first time. One of the lander's deflated airbags can be seen at lower right.

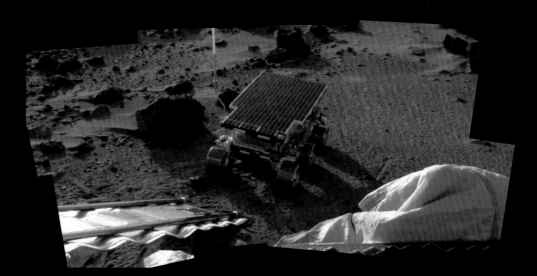

up a profile of the Martian atmosphere), and a meteorology package for measuring the surface weather conditions.

But the mission's most ambitious element, and one that captured the public imagination, was the Sojourner rover. This compact six-wheeled vehicle —named in honor of 19th century antislavery and women's rights activist Sojourner Truth following a nationwide schools competition—was just 26 inches (65 centimeters) long. Topped by a flat solar panel, the rover carried its own set of cameras, a prototype system for automatic hazard detection, a simple wheel for grinding rock samples, and the Alpha Proton X-ray Spectrometer (APXS), a "sniffer" for analyzing the chemical content of rocks.

Sojourner's maximum speed was just 0.4 inch (1 centimeter) per second, and it was only planned to operate for a few days—perhaps a month at best. In reality, though, it hugely exceeded expectations, trundling across the Martian surface for almost three months and covering a total distance of more than 330 feet (100 meters), though never straying more than 39 feet (12 meters) from the Pathfinder

base station, which it relied on for communications with Earth. In total, Sojourner sent back well over 500 photographs of the surrounding terrain, as well as analyzing 16 separate rock samples.

Pathfinder landed in Ares Vallis, an outflow channel in Chryse Planitia (see pages 126 and 156), not too far from the landing site of Viking 1, and mission scientists hoped that it would find conclusive evidence for water in the Martian past. The surrounding rocks, proved to be largely volcanic in nature, but one at least, known as "Yogi," had a smooth surface that suggested past erosion and transport by water. By the time communications with Mars Pathfinder were lost on September 27, the lander had returned more than 16,000 invaluable images of its environment, and the plucky little rover had, appropriately enough, shown the way for the more ambitious missions that would follow.

← This enhanced image reveals color differences among the rocks of Pathfinder's landing site. In the distance, the Sojourner rover is inspecting a rock nicknamed "Yogi," which proved to be a primitive volcanic basalt. Note the bright white material excavated by Sojourner's wheels as it turned a full circle—possibly similar to the silicates later unearthed by the Spirit rover (see page 188).
↓ The Pathfinder base station captured the component images of this 360-degree panorama during Sols 8, 9, and 10 of its mission. Two hills in the distance were soon nicknamed "Twin Peaks," and lie about 1.2 miles (2 kilometers) away from the landing site. Sojourner and the large rock Yogi lie just to the right of center.

Mars Global Surveyor

Mars Global Surveyor took advantage of the same launch window as Mars Pathfinder, afforded by the close approach of Earth and Mars in 1997. This spacecraft, however, was to be an orbiter—the first to take a new generation of imaging technology to the red planet.

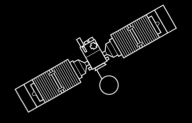

Mars Global Surveyor

In the two decades since the design of the Viking missions, studies of our own planet had been transformed by remote-sensing technology: cameras and other instruments aboard Earth-orbiting satellites that make measurements that are impossible from the surface, as well as providing us with a broader perspective on the planet. Within its boxlike body, Surveyor carried five main instruments, most of which shared some common principles with devices used closer to home.

The Mars Orbiter Camera (MOC) had originally been designed by Malin Space Science Systems to fly aboard the failed Mars Observer mission of 1992—it combined a monochrome, narrow-angle camera capable of resolving down to 56 inches (1.4 meters) per pixel from orbit, and two wide-angle cameras imaging the surface at red and blue wavelengths simultaneously at a resolution of 754 feet (230 meters) per pixel. Combining the wide-angle images allowed full-color images to be reconstructed. The Mars Orbiter Laser Altimeter (MOLA), meanwhile, was a device that used pulses from an infrared laser in a way analogous to radar to find

the precise altitude of the surface below. Firing 10 pulses per second for almost four years, its measurements allowed mission scientist on Earth to build up a detailed map of Martian topography. The Thermal Emission Spectrometer provided a means of mapping the chemical and mineral composition of the surface, while a magnetometer charted the planet's magnetic field in new detail.

Surveyor slipped into a highly elliptical orbit around Mars on September 12, 1997 (despite launching before Pathfinder, its much heavier mass of 2,266 pounds (1,030 kg) meant it gained less speed and took longer to reach its target). Like its sibling on the surface, Surveyor was tasked not only with investigating Mars, but also with testing new technologies and techniques, in this case "aerobraking," a means of adjusting the spacecraft's orbit without burning fuel by repeatedly skimming the top of the Martian atmosphere. This entire process took some 18 months after arrival, but eventually put Surveyor into its optimal orbit for imaging the surface, orbiting the planet once every two hours at an average altitude of around

← Mars Global Surveyor's basic structure consisted of two rectangular boxes on top of one another—one containing the rocket engines and propellant tanks, the other housing the spacecraft's computers, control systems and scientific instruments. A pair of 11.5-foot (3.5-meter) solar panels extended to either side of the probe, with steerable "drag flaps" on the ends to assist the aerobraking process.

250 miles (400 kilometers). The orbit was tilted at a high inclination, so that the probe would pass over both poles, and the period was chosen in order to remain "sun-synchronous"—in other words, Mars, Surveyor, and Sun would always be in the same relative positions and surface features would therefore all be illuminated from the same angle. As the satellite swept from pole to pole and Mars rotated slowly beneath its path, almost every area of the surface would be available for imaging at some point.

Throughout almost a decade of operation, Mars Global Surveyor sent back more than 240,000 MOC images, chronicling changes to the landscape across almost five Martian years. Its detailed

views of the surface revealed the unmistakable traces of recent liquid water flowing in areas such as Gorgonum Chaos, ancient sedimentary rocks in regions such as Becquerel crater and complex patterns in the polar ice caps (see pages 40, 98, and 106), and generated a huge list of intriguing targets for future orbiter observations and landers. Surveyor also transformed our view of Mars as an active planet, revealing the existence of the Martian dust devils, recent volcanism, landslides, and hints of long-term climate change (see pages 66 and 70). The mission finally lost contact with Earth in November 2006, following a problem with a software upgrade, but by that time it had already been joined by two successors—the 2001 Mars Odyssey mission and the Mars Reconnaissance Orbiter.

2001 Mars Odyssey

The Mars Odyssey orbiter took new remote-sensing technology to Mars and investigated its surface in unprecedented ways, making important discoveries along the way to becoming the longest-operating Mars space probe so far.

2001 Mars Odyssey

Despite their apparent success in breaking the run of bad luck associated with Martian space probes, the period following the successful arrival of Mars Pathfinder and Mars Global Surveyor saw the "curse of Mars" return with a vengeance. The Russian Mars '96 mission, launched just a couple of weeks after Surveyor, did not make it out of Earth orbit, and the 1998–9 launch window saw a string of failures —Japan's Nozomi spacecraft failed to reach its correct trajectory after running out of fuel, while in September, NASA's ambitious Mars Climate Orbiter succumbed to one of the most embarrassing failures in the entire history of space flight when a computer mix-up between metric and imperial units caused it to plunge into the upper atmosphere instead of entering orbit. Ten weeks later, in early December, the Mars Polar Lander mission also failed during its descent to the Planum Australe region bordering the southern ice cap.

A great deal therefore rested on NASA's solo 2001 mission—named Mars Odyssey in tribute to the famous science fiction novel and movie— and fortunately, it proved to be a huge success. Launched on April 7 and inserted into Martian orbit on October 24, Odyssey was designed to complement rather than supersede Mars Global Surveyor, and carried a suite of remote-sensing instruments that would allow detailed investigation of the thermal and chemical properties of the planet's surface. These instruments were the Thermal Emission Imaging System (THEMIS), a Russian-built Gamma-Ray Spectrometer (GRS) and the Mars Radiation Environment Experiment (MARIE).

THEMIS was a visible-light and infrared camera designed to image the landscape at eight different wavelengths—a sophisticated way of both distinguishing between mineral deposits that look identical in visible light and peering through surface dust layers to map the thermal properties of the underlying bedrock. By filtering out effects caused by daily temperature variations, THEMIS could reveal the terrain's characteristic "emissivity spectrum" —a mineralogical fingerprint that allowed mission scientists to pinpoint different rock types ranging from volcanic basalts and granites to sedimentary carbonate and silicate deposits. Images taken at a ninth, slightly longer wavelength, meanwhile, allowed THEMIS to study temperature variations in the Martian atmosphere.

Odyssey's Gamma-Ray Spectrometer, meanwhile, was a chemical "sniffer" designed to look for radiation emitted off the surface due to the impact of cosmic rays (high-speed particles from the Sun and the depths of space). Different elements emit characteristic wavelengths of high-energy gamma rays in response to these impacts, so the GRS was able to map the distribution of specific chemical elements in the Martian soil. MARIE was an experiment to measure the rate at which cosmic rays and high-energy radiation bombarded the spacecraft itself. Both can damage delicate electronics and living tissue, so understanding the radiation environment around Mars is an important factor in planning for any future manned missions.

It was GRS that made perhaps Odyssey's most important discovery, announced in May 2002. This was the confirmation of huge amounts of hydrogen within the Martian soil at both high and middle latitudes. Such enormous quantities of this lightweight element could only be explained by the presence of huge amounts of frozen water ice just below the surface. While the confirmation of present-day frozen water on Mars had long been anticipated, the sheer amounts involved came as a surprise to almost everyone. As William Boynton of the University of Arizona, principal investigator on the GRS instrument, put it at the time, "it may be better to characterize this layer as dirty ice rather than dirt containing ice."

Odyssey has proved to be an immensely durable spacecraft, overtaking Surveyor's previous record for the longest period of operation in Martian orbit in December 2010, and still operational at the time of writing. While the GRS instrument was decommissioned following a change in orbit and mission priorities during 2008, THEMIS continues to map the Red Planet and reveal the secrets of its mineralogy and weather.

↓ The most striking feature of Odyssey's design is the extended boom holding the 67-pound (30.5-kg) Gamma-Ray Spectrometer instrument. This keeps the sensitive germanium crystal that forms the core detector a safe distance from gamma rays produced by the spacecraft itself.

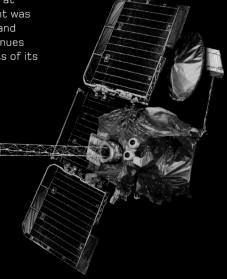

↑ Some of Odyssey's most spectacular and colorful results have come from its Thermal Emission Spectrometer (TES) instrument. False-color images such as this one of Meridiani Planum (see page 144) combine daylight views with nighttime infrared data to reveal variations in surface materials. Areas that remain warm after sunset (shown orange and red) are likely to be composed of denser and more solid rocks, while those that cool down rapidly (shown in blue) are thought to be drifts of fine-grained sand less capable of retaining heat.

← This TES view of a junction in the Noctis Labyrinthus (see page 118) shows differences in the surface including the valley's warm exposed rocky floor, looser, cool material associated with landslides on the canyon walls, and varied temperatures (indicating different amounts of dust) on the flat tops of the surrounding mesas.

Mars Express

The 2003 opposition of Mars saw the planet make its closest approach to Earth in 60,000 years—a unique opportunity that saw a veritable armada of spacecraft setting off for Mars, spearheaded by the European Space Agency's ambitious Mars Express mission.

Mars Express

Although this was Europe's first attempt to launch a Mars mission of its own, it had contributed much of the instrumentation to the failed Russian Mars '96 mission, and the Mars Express project drew on this experience to design and launch the mission in a very short timeframe. The spacecraft was launched aboard a Russian Soyuz rocket from Baikonur Cosmodrome in Kazakhstan on June 2, 2003 and arrived in orbit around Mars on Christmas Day. Mars Express consisted of two elements—the main orbiter spacecraft and a compact British-designed lander called Beagle 2. Named after the famous sailing ship that carried naturalist Darwin on his 19th-century voyage of discovery, the lander carried a range of experiments specifically designed to search for signs of past or present life on Mars. The bowl-shaped vehicle was just 40 inches (1 meter) across, with a total mass of 73 pounds (33 kg)—light enough to risk a descent to the Martian surface by a mixture of aerobraking, parachute, and protective airbags, without the need for cumbersome retrorockets. Sadly, however, while deployment from its mothership went according to plan, Beagle 2 never established communications from the Martian surface, and its fate remains unclear.

Fortunately, despite this initial disappointment, Mars Express itself was to prove a hugely successful mission. It carried a broad array of instruments including a mineralogical mapping spectrometer and others for studying the atmosphere, a radar altimeter known as MARSIS for studying conditions just below the visible surface, experiments for monitoring the space environment around Mars, and the High-Resolution Stereo Camera.

This last instrument, known by the initials HRSC, has delivered some of the most instantly impressive results, mapping the entire planet in full color down to 33-feet (10-meter) resolution and selected areas down to just 80 inches (2 meters) per pixel. What's more, an ingenious operating method allows HRSC to capture the same area from different angles in a single pass, producing a stereoscopic or 3D view that allows information about the shape of the surface to be extracted from the images. The camera has quite literally offered us a new perspective on the Martian landscape, revealing surface features in extraordinary detail. In 2006, for instance, the camera was used to produce three-dimensional images of the infamous "Face on Mars" (see page 124), demonstrating conclusively that it is nothing more than a trick of light and shadow. By the end of 2013, HRSC had completed a global map of Martian topography in unprecedented detail.

The Visible and Infrared Mineralogical Mapping Spectrometer (known by the French acronym OMEGA), meanwhile, studies the infrared emission of individual 330-foot (100-meter) squares of the surface in order to work out the likely minerals present. Almost as soon as it had arrived, Mars Express confirmed the presence of water in the Martian south polar cap, and it has subsequently compiled a complete geological map of Mars that will be used to guide future exploration.

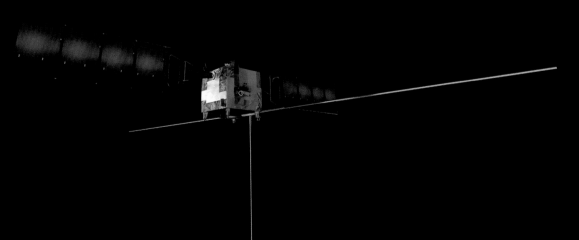

← Mars Express is a boxlike spacecraft with dimensions of just 4 foot 8 inches x 5 foot x 6 foot (1.4 x 1.5 x 1.8 meters). Three long booms extending around the spacecraft act as antennae for MARSIS (the Mars Advanced Radar for Subsurface and Ionospheric Sounding). The probe's unusual elliptical orbit brings it within 186 miles (300 kilometers) of Mars at its closest point, but takes it out to around 6,273 miles (10,100 kilometers) at its furthest, when the spacecraft turns away from Mars to communicate with Earth.

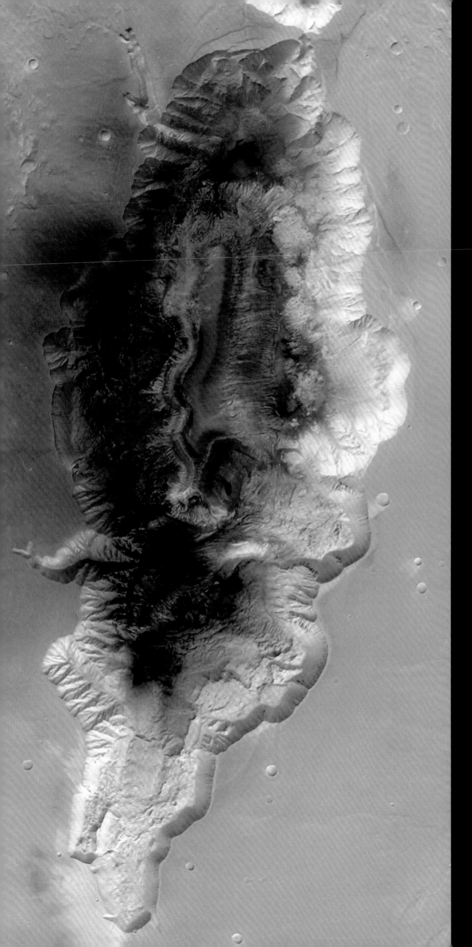

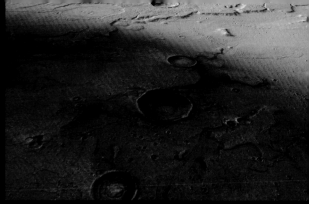

↑↑ By imaging the same area of the surface from two different directions, Mars Express's HRSC camera can generate topographic information as well as high-resolution images. Together, the two types of data can be used to produce stunning and enlightening perspective views of the surface, such as this landscape in Acidalia Planitia that shows signs of sculpting by water.

↑ This artist's impression depicts the ill-fated Beagle 2 lander on the Martian surface. Sadly, no signal was detected following its expected landing on December 25, 2003.

← HRSC can also simply be used to view the surface at extremely high resolutions, showing details down to around 5 feet (1.5 meters) per pixel at the lowest points in its orbit. The view at left shows Hebes Chasma, a deep trough within the Valles Marineris system (see page 84), while the image overleaf focuses on Nereidum Montes, a glacier-scarred highland region on the northern edge of the Argyre impact basin (see page 154).

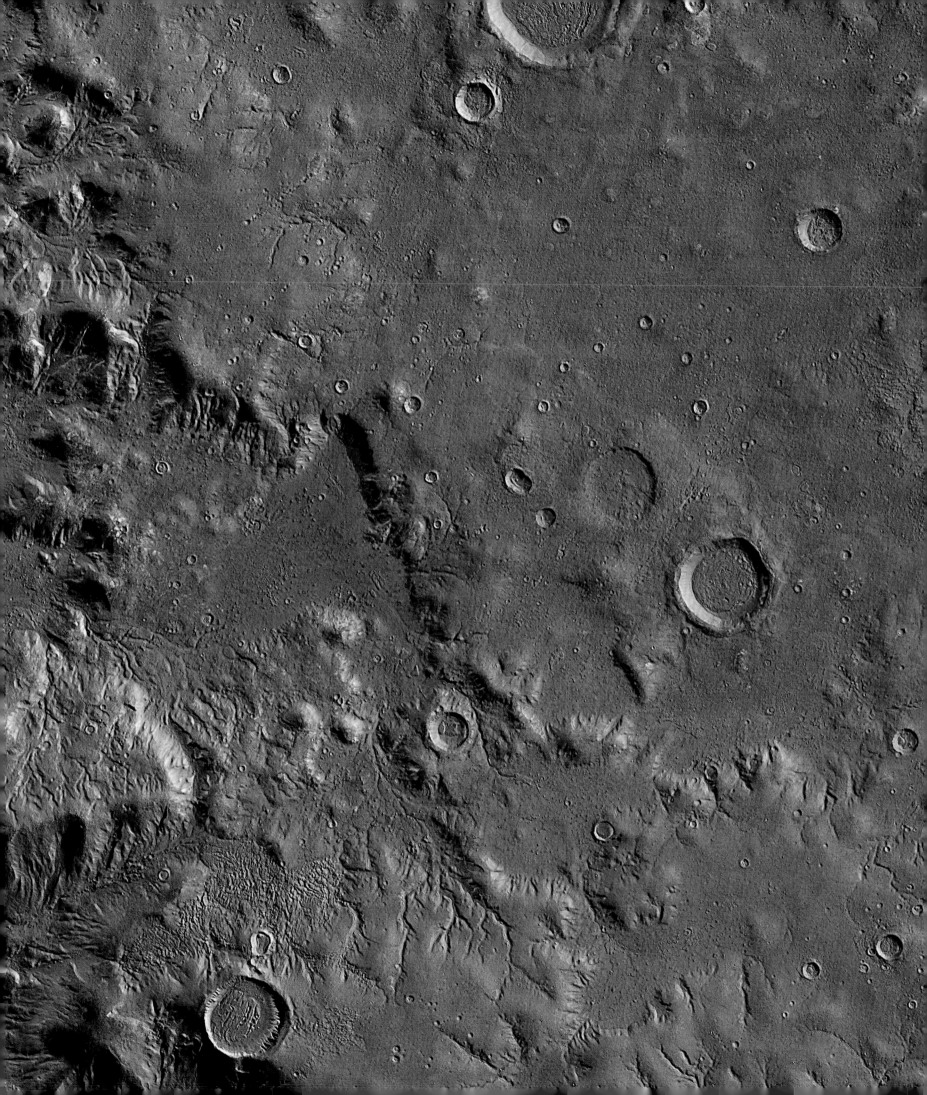

Mars Exploration Rovers

Following the success of the Sojourner Rover (see page 176), the obvious next step was to land a much larger and more ambitious vehicle on the surface of Mars—an autonomous rover that could function without a relay station, covering larger distances and operating for many months.

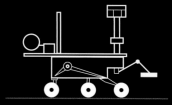

Spirit/Opportunity

In a throwback to the "belt and braces" policy of the Mariner and Viking era, NASA ultimately elected to dispatch two such rovers to Mars during the 2003 launch window. Officially classed as Mars Exploration Rovers A and B (MER-A and MER-B), they are better known as Spirit and Opportunity—names that were chosen following a student essay competition.

The two identical rovers were launched a month apart in the summer of 2003, and each took around seven months to make the journey to Mars. Their descent to the surface involved similar techniques to those used by Mars Pathfinder six and a half years before. An aeroshell covered in ablative material (heat-resistant material that chars and breaks away, carrying heat with it) protected the spacecraft during its initial plunge through the atmosphere, then a braking parachute slowed it down, before the shell separated and the lander unit was lowered down a reinforced bridle. Retrorockets in the backshell fired a short way above the surface to bring the descent to a stop, before the lander, encased in protective airbags, was dropped to the surface. Once they had stopped bouncing across the surface, the airbags deflated in sequence to leave the tetrahedral lander unit upright on the surface. Finally, the lander folded down its triangular sides and allowed the rover encased within to roll out onto the Martian landscape.

Each rover was about the size of a very small car, 5 feet (1.5 meters) tall, 7.5 feet (2.3 meters) wide, and 5.3 feet (1.6 meters) long—but with a total weight of just 396 pounds (180 kg). The rovers moved on six robust wheels, each with its own electric motors, independent suspension to help cope with rough terrain and independent steering that allowed the rover to make extremely nimble turns. They had a maximum speed of up to 2 inches (5 centimeters) per second, but normally moved much more slowly than this. Hazard avoidance software caused the rover to stop every 10 seconds, spending about 20 seconds analyzing images from four corner-mounted "Hazcam" cameras before choosing the safest route to proceed. This degree of autonomous steering, allowing the rover to make decisions for itself along a general route chosen by its human controllers, was essential in a situation where instructions take anything between 4 and 24 minutes to travel between Earth and Mars.

Three major instruments were supported at the top of the rover's "head," while others were mounted on the end of an articulated robot arm. The fixed

instruments were the full-color Panoramic Camera (Pancam), the monochrome, higher-resolution Navigation Camera (Navcam), and the Miniature Thermal Emission Spectrometer (Mini-TES), an infrared spectrometer capable of imaging the landscape at 167 distinct infrared wavelengths, enabling it to analyze the mineral content of rocks from a distance. The arm-mounted instruments, meanwhile, ranged from cameras, an optical microscope, and magnets for gathering samples of magnetic dust, to a pair of advanced spectrometers that bombarded sample rocks with different types of radiation to analyze their behavior and composition. The arm was also fitted with a rock abrasion tool for scraping away dust-covered or chemically altered layers and exposing deeper rocks to study.

Communication, meanwhile, could be maintained directly with Earth using a dish-shaped high-gain antenna (capable of transmitting data at high rates) or an omnidirectional low-gain antenna (sending data at lower rates). The low-gain antenna could also be used to send signals via orbiting spacecraft such as Mars Global Surveyor and Mars Odyssey. Power for each rover was supplied by 14 square foot (1.3 sq m) of solar panels, mounted in a winglike configuration and capable of a peak output of around 140 watts for about four hours around the middle of each Martian sol. This energy not only had to power the rover during that period, but also charge up the lithium ion batteries that would keep it operational at night. The only other energy supply came from eight 1-watt radioisotope heater units (RHUs)—capsules containing small amounts of radioactive material that were used to supply a small amount of heat to critical rover systems. These allowed the rovers to withstand temperatures between −40°F (−40°C) and 104°F (+40°C).

↑ Two artists' impressions capture stages in a Mars Exploration Rover's landing. At left, the backshell retrorockets fire to slow the descent of the balloon-encased lander. At right, the lander, now cut loose from its tether, bounces toward its eventual landing site on the Martian surface.

→ This artist's conception shows a Mars Exploration Rover in operation on the Red Planet. Key features include the main navigation and panoramic cameras mounted on the central pillar, and the extended arm carrying the microscopic imager, rock abrasion tool, and spectrometers. A dish-shaped high-gain antenna and upright low-gain antenna are both mounted on the upper surface.

Spirit

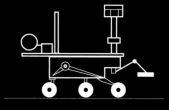

Spirit

NASA's Spirit rover touched down on Mars on January 4, 2004, near the middle of the 103-mile (166-km) Gusev crater. A large canyon system, Ma'adim Valles, opens into Gusev, and orbital images raised hopes that the crater floor might be a dried-up lakebed filled with sedimentary rock.

Spirit's landing area, named the Columbia Memorial Station in honor of the seven astronauts who had lost their lives in the Space Shuttle disaster of February 2003, was a rolling landscape littered with small rocks, with a series of low hills (soon named after Columbia's individual crew members) in the distance. Spirit's controllers on Earth determined that these were the most likely place to find interesting geology, and so the rover set out on a long journey to Husband Hill, with occasional pauses to investigate some of the rocky debris in its path (they turned out to be volcanic in origin, but showed possible traces of water-laden mineral deposits within them). Spirit had already outlasted its 90-sol primary mission by the time it reached the foot of the Columbia Hills on Sol 159, where it began to investigate interesting rocks and soon identified the presence of hematite (a mineral that, at least on Earth, forms in the presence of water).

Spirit spent much of 2005 climbing up Husband Hill and investigating the rocks at its summit. One unexpected highlight was the chance to photograph dust devils passing nearby on the crater floor. At a location called Home Plate, it identified rocks that were notably rich in carbonate minerals, which form in wet conditions but dissolve rapidly in acid. In early 2006, the rover drove to nearby McCool Hill, where it was planned that it could sit out the coming Martian winter on a south-facing slope. Along the way, one of Spirit's front wheels, which had been giving trouble for some time, finally stopped working altogether. The rover was able to continue on its route by the simple expedient of turning around and driving backward. The faulty wheel turned out to be a mixed blessing—dragged behind the rover it scraped away at the surface to reveal the material that lay below, and in March 2007, exposed a vein of distinctive bright dust. When the rover investigated, this turned out to be rich in silica (silicon dioxide), and mission scientists concluded that it had almost certainly been deposited around a hot spring—the kind of environment that, on Earth at least, is ideal for microbial life.

By the summer of 2007, Spirit had returned to the vicinity of Home Plate, but now both it and Opportunity were threatened by global dust storms that swept across Mars. As atmospheric haze blocked out much of the sunlight for two months or more, power output from the rovers' solar panels fell dangerously low. After spending some months in hibernation mode to preserve the charge in its batteries, Spirit limped onto a south-facing slope in

early 2008 to prepare for another winter. Normal operations could not resume until almost a year later, when winds blew away much of the dust that had accumulated on Spirit's solar panels, and the rover was able to risk long-distance drives once again.

Sadly, though, Spirit soon ran into more trouble, finding itself caught in a "sand trap" and unable to move at the start of May 2009 (Sol 1892). Unable to gain traction from the wheels, engineers back on Earth tried to use mock-ups and find a way to free the rover, but eventually had to concede defeat— Spirit was firmly bedded in at a location named Troy. With no way of preparing itself for the onset of the 2010 winter season, Spirit continued scientific work from its now-stationary base until early 2010, when its power supply finally failed—its last communication with Earth came on March 22, 2010 (Sol 2010).

Spirit's most surprising finding was that Gusev crater's geology showed little sign that it had ever been the sedimentary lakebed previously suspected. Instead, the rocks and soil that littered its surface were largely volcanic in nature. Weighing Spirit's samples against the evidence of large-scale lake features in satellite photographs, most experts have concluded that Gusev is indeed an ancient crater lake—just one that was subsequently buried under a layer of volcanic rock. The discovery of the carbonate and "hot spring" deposits in the Columbia Hills was undoubtedly a boost for the idea that conditions here might once have been suitable for life, but more conclusive evidence for a wet Martian past would come from Spirit's sibling.

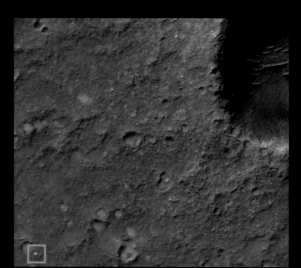

→ This unusual self-portrait of Spirit, compiled from a mosaic of images in November 2008, shows the rover's solar panels still gleaming in the cold Martian sunlight. Cleaning events that periodically sweep away accumulated dust allowed both of the Mars Exploration Rovers to greatly exceed their planned operational lifetimes.

← Spirit's three-petaled landing platform is visible as a bright spot at bottom left in this Mars Reconnaissance Orbiter image, covering an area about 1,600 feet (500 meters) across. At upper right is Bonneville, the closest large crater to the landing site.

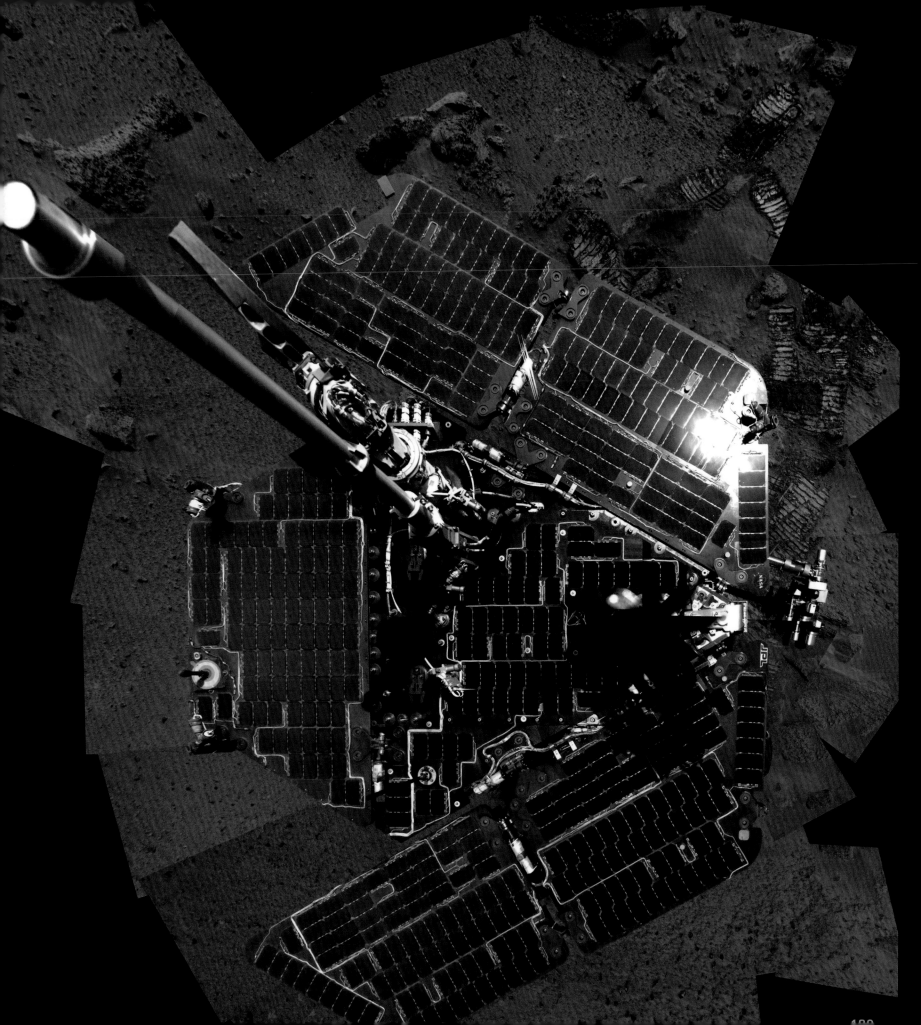

A faint, distant sun sets over the rim of Gusev crater in this spectacular photograph, captured by Spirit around 18.07 local time on Sol 489 of its mission. Beautiful images like this not only provide an idea of what it would be like to walk on the surface of the Red Planet, but also offer valuable scientific information about the distribution of dust in the Martian atmosphere.

Opportunity

Launched from Earth a month after Spirit, Opportunity landed three weeks after its sibling, on January 25, 2004. Targeting the Meridiani Planum region, it made a series of stunning discoveries in a mission that lasted some 30 times longer than initially planned.

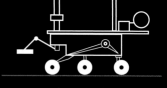

Opportunity

Shielded by its protective cluster of airbags, Opportunity landed some 16 miles (25 km) from its intended landing site, bouncing to a halt at the center of a small 72-foot (22-meter) depression that was subsequently named Eagle crater. As the rover deployed its solar panels to charge in the weak Martian sunlight, and its cameras took their first panoramic images of its surroundings, scientists back on Earth were delighted to find they had scored a "hole in one," landing in a location where the deeper layers of the Martian soil were naturally exposed. Exposed outcrops in the crater walls clearly showed thinly layered sedimentary rocks, with a mix of reddish Martian dust and coarse but light gray material. These layered deposits, laid down over time and later compressed from above, must have formed from either volcanic ash, or wind-blown or water-lain deposits. Close-up studies of the rock layers during the first weeks of Opportunity's mission revealed evidence that suggested the presence of water during their formation.

Shortly afterward, Opportunity exited Eagle crater without much difficulty, making its way toward the much larger Endurance crater, an intriguing target that had been identified from above by the orbiting Mars Global Surveyor. After skirting the edge, the scientists decided that a descent into Endurance to investigate the rocks exposed around its edges was worth the risk, even if Opportunity might not make it out again. Ultimately, the rover spent 180 Martian sols in the crater before moving on.

Early in 2005, the rover came across an unusual rock lying close to its own discarded heat shield. This proved to be the first meteorite identified on the surface of another planet. A few months later, as it made its way south to new targets, it became dangerously stuck in a sand trap that engineers back on Earth nicknamed Purgatory dune. Escaping took more than a month, though this was partly because the mission planners carefully rehearsed various escape plans on Earth before giving the rover its instructions.

Opportunity took more than a year to reach its next major target, a much larger crater called Victoria, some 4.3 miles (7 kilometers) from its landing site. With a diameter of some 2,400 feet (730 meters), Victoria has high rock outcrops along its walls, exposing deep layers of rock from the planet's interior. Opportunity arrived in September 2006 and began to make its way around the crater rim, sending back images that revealed a deep field of dunes at its center. Perched on the edge of the crater, the rover was photographed from space by the newly arrived Mars Reconnaissance Orbiter.

From April 2007, a series of "cleaning events" thought to be associated with the ubiquitous Martian dust devils helped to remove accumulated dust from Opportunity's solar panels, restoring its power output to near-maximum levels. By now, scientists had identified a route to take the rover down into the crater, but before they

↓ On its way to Victoria crater (see page 128), Opportunity explored a smaller crater called Erebus in late 2005—this panoramic image from the crater rim consists of no fewer than 635 separate PanCam images taken through four different filters. However, although the crater is almost 1,150 feet (350 meters) across, it is very shallow.

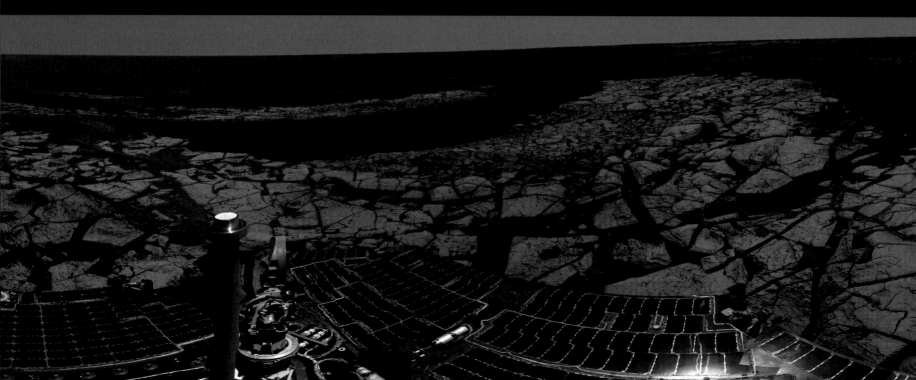

could begin the descent in earnest, a global dust storm engulfed both rovers, blocking out almost 99 percent of the sunlight reaching Opportunity's panels, starving it of energy and forcing it to draw power from its onboard batteries. All nonessential systems were shut down, and throughout July 2007, NASA engineers held their breath, hoping that battery levels would not fall low enough to trip a low-power fault (a failsafe system from which the rover would attempt to recover, but with no guarantee of success). Fortunately, the storm eased in early August, the rover's batteries began to charge and, within weeks, Opportunity was able to resume normal operations.

After a detailed investigation of Victoria, Opportunity set off again on an epic 12-mile (19-kilometer) trek to an even larger crater, the 14-mile (22-kilometer) Endeavour. Along the way, it made numerous diversions to investigate other craters and several interesting rocks: some of these proved to be meteorites, while another seemed to be a large chunk of ejecta, flung out from deep within the Martian crust by a significant impact.

Opportunity finally arrived at the rim of Endeavour on August 9, 2011. Even without entering the crater, the rover has been able to make important discoveries along the edge, including ejecta produced during the crater's formation and new types of rock not seen elsewhere on Mars. One of the most significant discoveries of all came in December 2011, when Opportunity confirmed that a rock formation called Homestake is formed from gypsum, a mineral that can only form in the presence of water. Mission scientists concluded that Homestake is possibly the remnant of an ancient spring that flowed through fissures in the Martian rock.

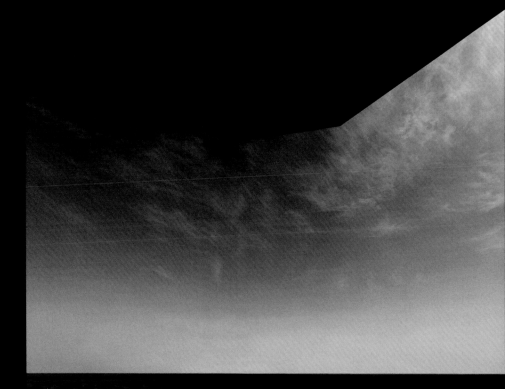

↑ This striking mosaic view highlights the presence of high-altitude glowing or "noctilucent" clouds. Such clouds, discovered on Mars for the first time by Opportunity, may help certain areas of the planet retain warmth at night. In the

← This synthetic view from the Mars Exploration Rover team shows Opportunity in context as it patroled the inner slopes of Endurance crater in June 2004. The image combines a computer-rendered model of the rover with a photographic mosaic gathered by Opportunity itself.

← A pair of images show the hematite-rich "blueberries" discovered by Opportunity around Eagle crater. The rock on the left, nicknamed "berry bowl" was notably rich in these small stones—the light circle shows where it was scrubbed clean with the rock abrasion tool for spectroscopic analysis. On the right, a microscopic view shows the blueberries in detail.

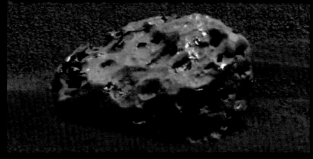

← This heavily pitted, soccer-sized rock, discovered by Opportunity in early 2005, proved to be the first meteorite every identified on the surface of another planet. Analysis by the rover's spectrometers confirmed that it is composed largely of iron and nickel—such meteorites are thought to be fragments from the core of large asteroids shattered to pieces by collisions in the asteroid belt.

← This eerie landscape, captured by Opportunity in March 2012, shows the view across Endeavour crater in the weak late afternoon sunlight of a Martian winter. The crater floor and far rim appear bluish thanks to the scattering of light by atmospheric particles (the same phenomenon that tints distant objects on Earth.

Mars Reconnaissance Orbiter

Launched in August 2005 as a successor to the Mars Global Surveyor mission, NASA's Mars Reconnaissance Orbiter (MRO) was equipped with onboard cameras capable of imaging the surface in more detail than ever, alongside a suite of advanced remote-sensing instruments.

Mars Reconnaissance Orbiter

The MRO mission had originally been studied and developed for possible launch in the 2003 launch window, but was ultimately postponed to make room for the twin Mars Exploration Rover missions. Instead, after a seven-month journey, MRO finally entered orbit around Mars on March 10, 2006. Following six months of aerobraking in which the probe skimmed the upper layers of the Martian atmosphere in order to adjust its altitude, a final burn of its onboard thrusters inserted it into a final orbit ranging between 196 and 155 miles (316 and 250 kilometers) above the surface. By this time, solar conjunction, in which Mars would pass behind the Sun and MRO would lose communications with Earth, was imminent, so the orbiter did not begin its full scientific mission until early November.

MRO's showpiece instrument is a camera known as HiRISE (the High Resolution Imaging Science Experiment), which has been likened to a microscope in space. In fact, it consists of a 20-inch (0.5-meter) telescope attached to a camera that images the surface at three wavelength bands (blue-green, red, and near-infrared). The resolution of the resulting images, around 12 inches (30 centimeters) per pixel, is equal to those of many Earth-orbiting satellites. HiRISE produces characteristically long, striplike images as it views the terrain passing beneath it, and it can also generate pairs of stereoscopic images that reveal the varying topography of the landscape. Lower-resolution cameras, meanwhile, provide wider-angle views of the surface in both grayscale and color.

The orbiter's other instruments include a visible and near-infrared spectrometer called CRISM, which splits images from the surface into 544 narrow channels with specific wavelengths in order to reconstruct a spectrum and reveal chemical properties of the rocks. Another spectrometer studies the Martian atmosphere by looking along MRO's orbit at a shallow angle. Finally, the Shallow Subsurface Radar instrument (SHARAD) is a ground-penetrating radar that fires beams of radio waves into the Martian surface and measures their reflections. It can detect distinct layers in the Martian soil less than 33 feet (10 metres) thick, down to depths of around 3,300 feet (1 kilometer), and was built by the same Italian team who provided the lower-resolution but more highly penetrating MARSIS radar carried aboard ESA's Mars Express (see page 182).

A more prosaic but equally important feature of MRO is its communications package, known as Electra. This is by far the most advanced telecommunications system ever sent into space, with a 10-foot (3-meter) high-gain antenna capable of sending data back to Earth ten times faster than previous Mars orbiters. This not only allows the spacecraft to transmit the enormous amounts of data generated by its cameras and other scientific instruments, but also lets it function as a communications relay station for landers and rovers on the planet's surface.

MRO has led to massive improvements in our understanding of Martian surface mineralogy and chemical composition, as well as returning stunning images of the surface. Its discoveries include the presence of ice beneath the northern plains and within the glacierlike lobate debris aprons, signs of Martian climate change such as the starburst features at the south pole, and evidence of standing bodies of water on ancient Mars. What's more, HiRISE's resolution is high enough to clearly photograph the Spirit, Opportunity, and Curiosity rovers as well as other Martian landers.

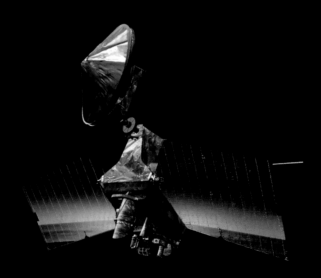

← Mars Reconnaissance Orbiter was equipped with twin solar panels, each with an area of 102 square feet (9.5 square meters), and a pair of rechargeable batteries for powering the spacecraft when it faces away from the Sun. No fewer than 20 control thrusters allow the spacecraft to adjust its orientation and orbital path.
→ Images from MRO's HiRISE camera typically exaggerate color differences on the Martian surface. Blue-gray tints in the soil are particularly noticeable, as seen in this image of Cerberus Fossae, an ancient fault that once released floods of lava across the nearby landscape.

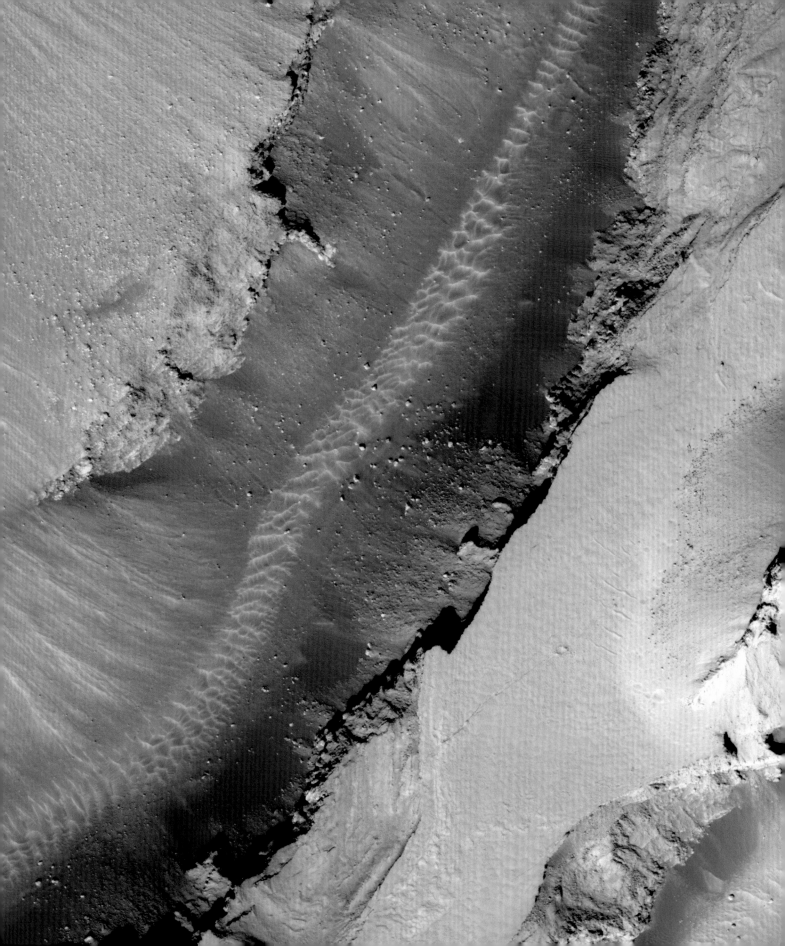

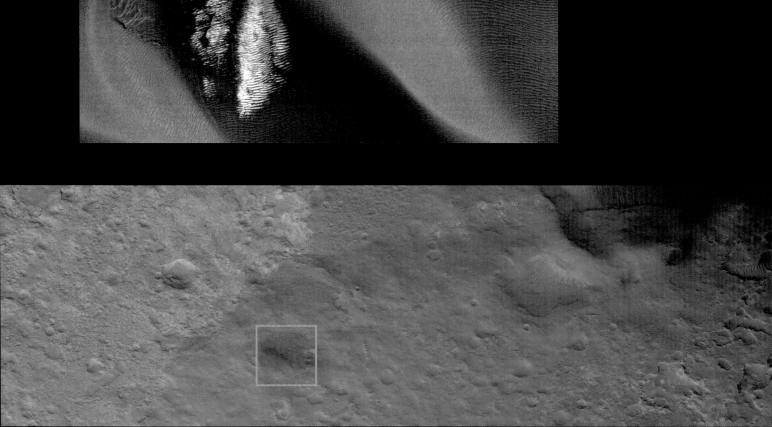

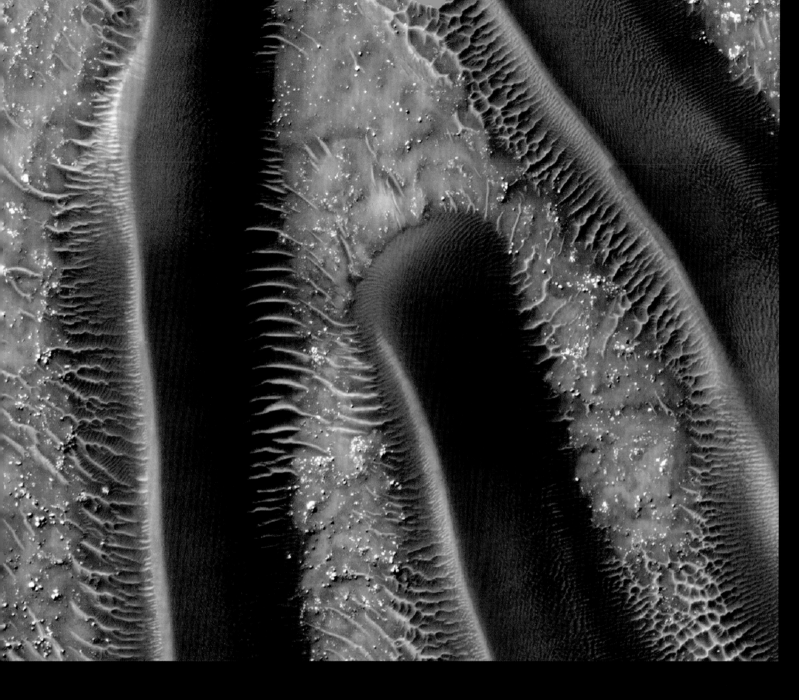

Phoenix

Phoenix

NASA considered a number of low-cost missions for the 2007 launch window, but the one it ultimately selected was designed to settle some unfinished business and resolve some key questions about the planet's polar caps.

The Phoenix lander, developed by the University of Arizona, made use of instruments designed for the failed Mars Polar Lander of 1999. Costs were also kept down by the use of the mothballed lander from a canceled mission known as Mars Surveyor 2001. But while Mars Polar Lander would have targeted the planet's south pole, Phoenix planned to study conditions around the north polar cap.

Nevertheless, similar challenges would need to be met—the lander would need a larger area of solar panels in order to draw enough power from the weaker high-latitude sunlight and the instruments would need to withstand colder conditions than those faced by equatorial missions. The first problem was addressed by the addition of a pair of dishlike solar panels, with a total area of 31.2 square feet (2.9 square meters), which unfolded and deployed after landing; the second, by the addition of a number of electrical heaters, which would prove to be a substantial drain on the lander's resources. Nevertheless, it was clear from the outset that Phoenix's mission would have a limited duration of necessity, since the spacecraft would be very

unlikely to withstand the harsh Martian winter. Phoenix was launched on a Delta II rocket from Cape Canaveral on August 4, 2007, and descended into the Martian atmosphere on May 25, 2008. A delay in the parachute opening sequence during descent led Phoenix to touch down a little way off target, in the Green Valley area of Mars's great northern plain, the Vastitas Borealis (see page 110). Mars Reconnaissance Orbiter snapped the lander suspended from its parachute during its final descent. Unlike Mars Pathfinder and the Mars Exploration Rovers, the final stages of landing were controlled using retrorockets mounted beneath the lander body itself, rather than bouncing onto the surface in protective airbags.

The mission was designed to confirm the presence of subsurface water ice, so strongly suggested by Mars Odyssey's Gamma-Ray Spectrometer results in 2002 (see page 44). Orbital images taken during the mission's planning stages had revealed polygonal patterns in the soil of the target area—patterns which on Earth are a likely sign of permafrost just beneath the surface. The descent imager mounted

↙ Phoenix's landing site in the Vastitas Borealis lay midway between the permanent icecap of the Planum Boreale and the northern slopes of the Tharsis rise around Alba Patera (see pages 98 and 116).
↓ An artist's impression shows Phoenix's descent to the Martian surface., supported on its retrorocket thrusters. This unusual landing method allowed the lander to make a precisely controlled landing on the surface so that its unfolded solar panels would be oriented along an east-west axis and able to harvest the maximum amount of precious solar energy.

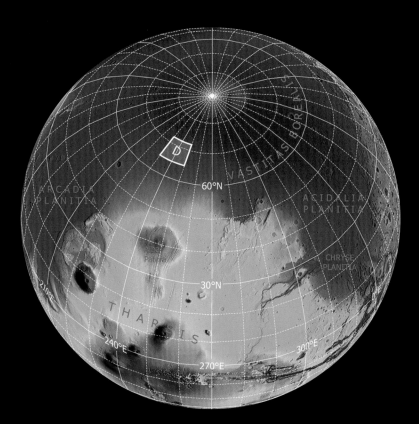

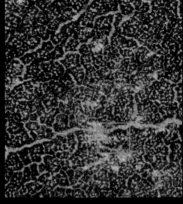

in the spacecraft's underside and the lander's pillar-mounted stereoscopic camera soon confirmed these patterns close up, and one unexpected consequence of this was that the retrorockets blew away the dust from some nearby areas of the surface, exposing bright underlying material that was soon identified as water ice.

The lander's other instruments included a weather station and a robot arm for photographing soil details and collecting samples for processing in the automated laboratories carried onboard. Experiments included the Thermal and Evolved Gas Analyzer (TEGA)—a combined furnace and mass spectrometer capable of heating soil samples and identifying the vapors released at different temperatures—and the Microscopy, Electrochemistry and Conductivity Analyzer (MECA), which studies various soil properties, including the soil's electrical and thermal conductivity, chemical reactions when wet, and microscopic structure.

The mission's discoveries not only confirmed the existence of substantial water ice in the Martian soil, but also found other chemicals such as carbonate minerals (likely remnants of a wetter Martian past) and perchlorate, a chemical that is generally thought of as inimical to most forms of microbial life.

Phoenix's primary mission was designed to run for 90 sols around the northern midsummer solstice (June 25, 2008)—a period when the Sun would remain permanently above the horizon from its latitude of 68° North. In the end, it operated for 157 sols before the onset of fall overcame it. The lander saw its first sunset in early September 2008, shortly after the end of its primary mission, and the rover went into a "safe mode" shutdown due to loss of power for the first time on October 28. After a brief revival to allow for a more orderly shutdown of systems, Phoenix sent its final message to Earth, the single word "TRIUMPH" in binary code, on November 2. Despite the spacecraft being encased in dry ice by the expanding polar cap through the ensuing winter, attempts were made to renew contact during the following Martian spring in January 2010, but these proved unsuccessful.

Curiosity

When the car-sized Curiosity rover touched down on Mars in August 2012, scientists hoped that it would answer a very precise question—could the ancient Martian environment have once supported life as we know it?

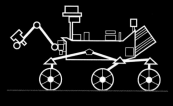

Curiosity

Launched from Cape Canaveral in November 2011, Curiosity was a major advance on its predecessors Spirit and Opportunity, not only technologically, but also in terms of sheer size and ambition. With a mass of 1,980 pounds (900 kg), the rover itself weighed almost five times as much as the previous Mars Exploration Rovers—so much that simply getting Curiosity onto the surface of Mars required a radical new approach. Instead of bouncing onto the surface encased in protective airbags and a protective pyramid, Curiosity was released fully exposed from its entry capsule following initial parachute braking, and was gently lowered to the ground on a 66-foot (20-meter) tether beneath a hovering, rocket-powered "sky crane." Touchdown (at a speed of less than 1 mile per hour or 1.6 kilometers per hour) triggered the automatic release of the sky crane, which then flew off to a safe distance before crashing.

After a lengthy discussion of potential landing sites, including detailed analysis of images from Mars Reconnaissance Orbiter and other spacecraft, mission planners elected to target Gale, a 96-mile (154-kilometer) impact crater just south of the Martian equator, with plentiful sedimentary rocks and other signs of a watery past (see page 112). Curiosity's ultimate destination is Aeolis Mons, the crater's 3.1-mile (5-kilometer) central peak, which is thought to be surrounded by sedimentary layers dating back through several billion years of Martian history.

Curiosity's design is very different from that of its smaller precursors, most obviously because the rover's larger mass and higher power demands led to the abandonment of solar energy as a primary power source. Instead, Curiosity carries a radioisotope thermoelectric generator (RTG)—a generator that produces electricity from the heat released in the decay of a sample of radioactive plutonium.

The rover's central mast carries its main cameras, including a remote-sensing instrument that can analyze rocks in its surroundings, while its robot arm carries a close-up camera and an X-ray spectrometer—a chemical "sniffer" for investigating rocks in detail. A drill and dust-collecting tool, meanwhile, allow it to extract rock samples for the first time, delivering it to two separate onboard laboratories—the Chemistry and Mineralogy suite (CheMin), and Sample Analysis at Mars (SAM). As its name suggests, CheMin focuses

on analyzing the chemistry and structure of the rock samples, while SAM is designed to look for organic molecules and gases.

Early studies of Curiosity's surroundings revealed that the probe had landed in the remains of a dried-up streambed, amid typically gravelly "conglomerate" rocks. Early analysis of dust and soil samples confirmed similarities between Martian dust and basaltic volcanic rocks found on Earth, and began to reveal complex and unexpected molecules in the Martian soil.

In February 2013, Curiosity collected its first sample of Mars rock from "John Klein rock" in a region called Yellowknife Bay, and struck something of a jackpot. The rock sample had traces of carbon, nitrogen, oxygen, phosphorous, and sulfur, all vital elements for life on Earth. In addition, the sample consisted of at least 20 percent clay minerals that could only have formed in fresh water. The presence of calcium sulfate mineral veins, confirmed that the water was neither acidic nor alkaline—in other words, it would have been a perfect environment for Earthlike life. As Curiosity scientist John Grotzinger put it, "we have found a habitable environment which is so benign and supportive of life that probably if this water was around, and you had been on the planet, you would have been able to drink it."

Since then, Curiosity has continued its drive to Aeolis Mons—an 5.3-mile (8.6-kilometer) journey that will take an estimated three years in total. It has continued to analyze rocks and soil samples along the way, with some significant discoveries including the presence of up to 3 percent water by weight in the local Martian soil, and the confirmation (through measurements of the rare atmospheric gas argon) that certain families of meteorite thought to come from Mars do indeed originate on the Red Planet. One of its most puzzling and disappointing results, however, has been the failure to find atmospheric methane in contradiction to previous results from Mars Express (see page 75). Doubtless many more discoveries will be made as Curiosity continues its exploratory trek.

→ Curiosity's robot-arm-mounted camera allowed it to produce this astonishing self-portrait, a mosaic of 55 separate high-resolution images. When this image was taken on Sol 84 (October 31, 2012), the rover had just finished collecting its first soil samples at a location named "Rocknest," and four scoops can be seen in the sand at lower left. Aeolis Mons rises in the background on the right, while the north wall of the crater can be seen further in the distance.

↓ An artist's impression shows Curiosity's final descent into Gale crater beneath its supporting sky crane. The successful trial of this landing technique should pave the way for even larger and more capable robot rovers to explore the Martian surface in the future, and may even be used by future manned missions.

➔ A number of instruments cluster at the top of Curiosity's Remote Sensing Mast. The white box with the large aperture is ChemCam, a device that can fire a laser beam to heat nearby rocks and analyze the results. Below this are the square apertures of MastCam, which combines a pair of wide-angle and narrow-angle cameras to produce high-resolution images of Curiosity's surroundings. Small circular lenses to either side are part of the rover's navigational system.

↓ Seen at the center in this view of Curiosity's robot arm "turret," the Mars Hand Lens Imager (MAHLI) is a specialized camera designed for imaging surface details and soil samples in visible and ultraviolet light. Capable of focusing at distances ranging from a few millimeters to infinity, The camera itself is surrounded by LED lamps to illuminate its subjects.

◢ Curiosity's arm turret acts as both a scientific instrument and an excavation tool. In this image, taken during the rover's first exploratory drilling attempt, the Powder Acquisition Drill System (PADS) is just touching the surface and the Dust Removal Tool, a system of rotating wire brushes, extends to the lower left.

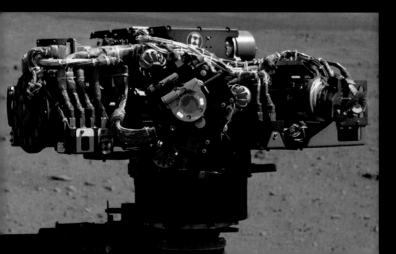

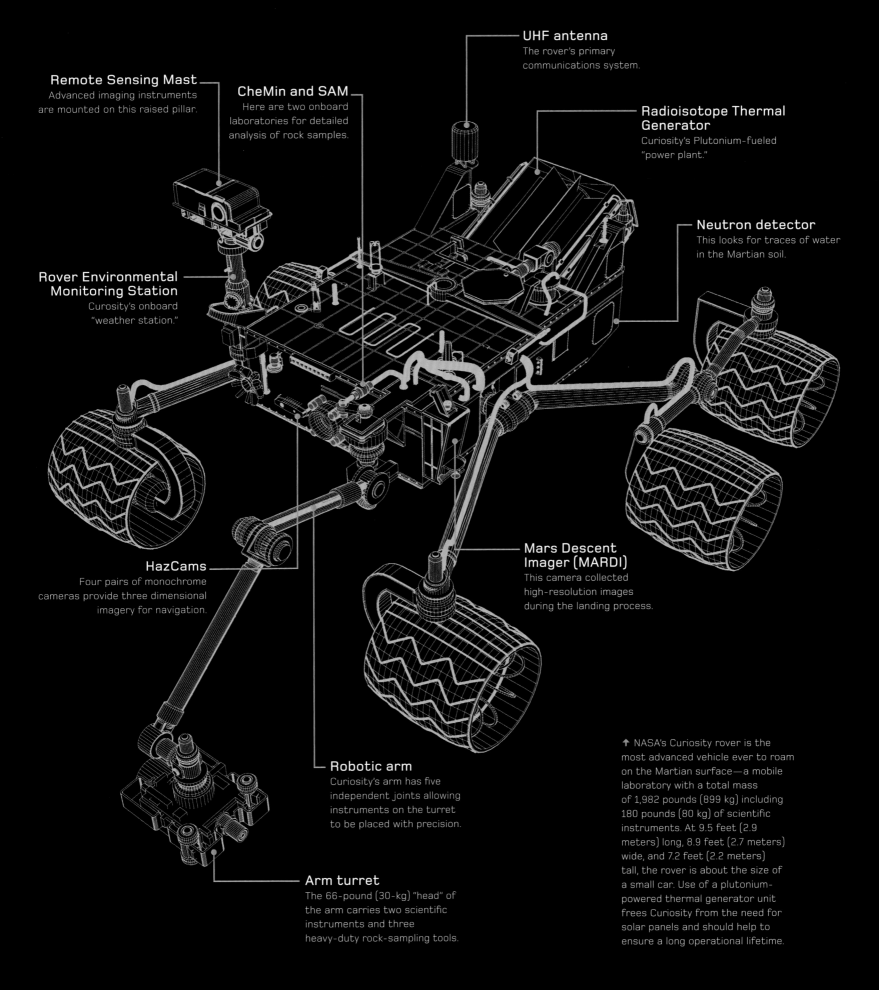

Remote Sensing Mast
Advanced imaging instruments are mounted on this raised pillar.

CheMin and SAM
Here are two onboard laboratories for detailed analysis of rock samples.

UHF antenna
The rover's primary communications system.

Radioisotope Thermal Generator
Curiosity's Plutonium-fueled "power plant."

Neutron detector
This looks for traces of water in the Martian soil.

Rover Environmental Monitoring Station
Curosity's onboard "weather station."

HazCams
Four pairs of monochrome cameras provide three dimensional imagery for navigation.

Mars Descent Imager (MARDI)
This camera collected high-resolution images during the landing process.

Robotic arm
Curiosity's arm has five independent joints allowing instruments on the turret to be placed with precision.

Arm turret
The 66-pound (30-kg) "head" of the arm carries two scientific instruments and three heavy-duty rock-sampling tools.

↑ NASA's Curiosity rover is the most advanced vehicle ever to roam on the Martian surface—a mobile laboratory with a total mass of 1,982 pounds (899 kg) including 180 pounds (80 kg) of scientific instruments. At 9.5 feet (2.9 meters) long, 8.9 feet (2.7 meters) wide, and 7.2 feet (2.2 meters) tall, the rover is about the size of a small car. Use of a plutonium-powered thermal generator unit frees Curiosity from the need for solar panels and should help to ensure a long operational lifetime.

mission (August 20, 2013), Curiosity paused on its long drive to capture the astonishing site of an annular solar eclipse as Phobos, the larger and closer Martian moon, swept across the disk of the Sun.

← This close-up view of Curiosity's own wheels was taken as part of a series of tests of the rover's robot arm and its MAHLI camera. Each wheel is 20 inches (50 centimeters) across, with independent suspension and gearing that allow it to maintain traction in tough Martian conditions. Photographs such as these allow engineers on Earth to assess the condition of the rover's various components, and hopefully keep it running for longer.

↓ Mission scientists named the rise on the right of this panoramic image "twin cairns" after the two grayish rocks at its peak. The view is a mosaic of seven separate images from Curiosity's mast-mounted camera, taken on Sol 343 (July 24, 2013). The image has been "white balanced"—treated to suggest its appearance under Earthlike lighting conditions—a procedure that can help to reveal significant differences in rock and soil color and composition.

← This view looks across the plains of Gale crater toward Curiosity's primary destination, Mount Sharp or Aeolis Mons. Richly structured layers are clearly visible running all the way up the side of the mountain, promising a treasure trove of sedimentary rocks for the rover to study. The image was taken during tests of the rover's 3.9-inch (100-millimeter) MastCam shortly after landing in August 2012, and shows foothills roughly 3.4 miles (5.5 kilometers) away, rising to a peak some 10 miles (16.2 kilometers).

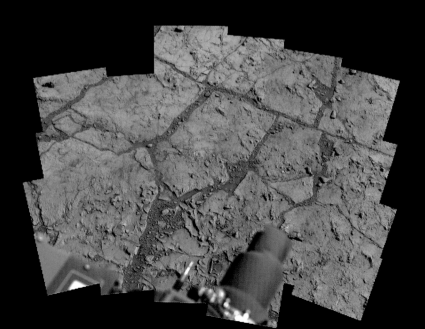

↑ This sunset panorama combines 40 images from Curiosity's MastCam captured on Sols 170 and 176, and subsequently processed by French Mars researcher Olivier de Goursac. The mages were all taken between 3 p.m. and 4 p.m. local solar time.

← A view from MastCam shows the context of John Klein rock (far left, at center), target for Curiosity's first exploratory drilling, while a close-up view from the MAHLI camera shows the aftermath of the drilling process, The hole was just 0.6 inches (1.6 centimeters) in diameter, and the gray dust it produced suggested that iron in the rock's interior is less oxidized than that on the rusty surface.

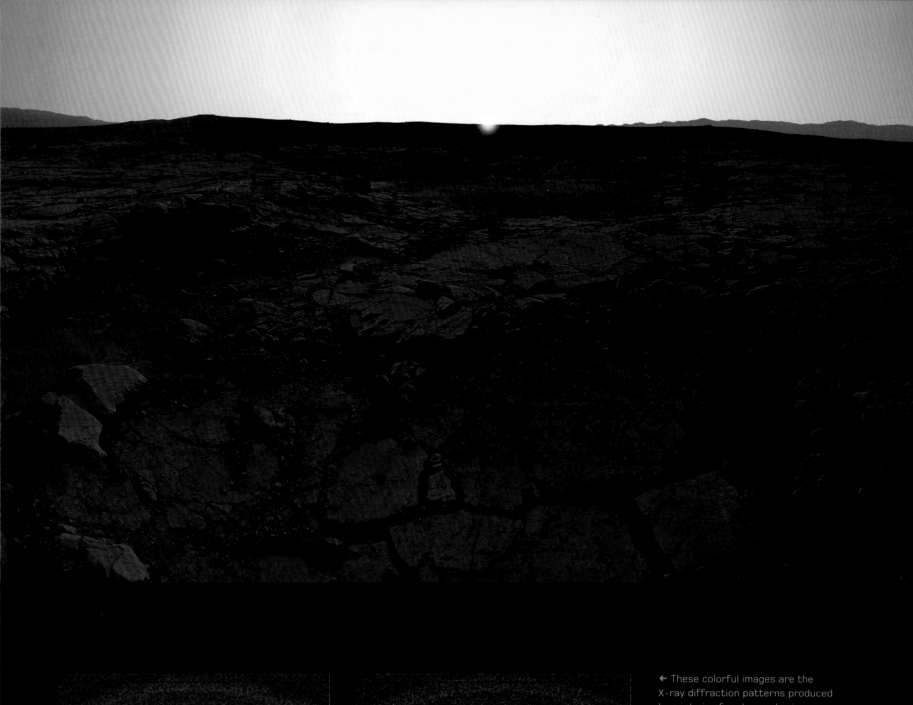

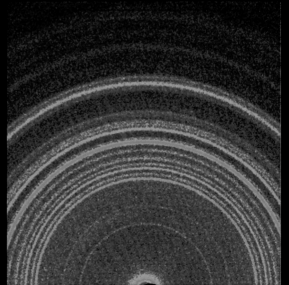

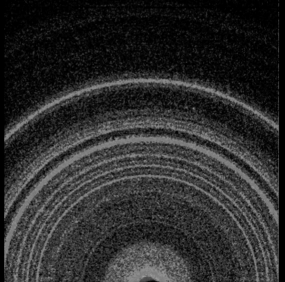

← These colorful images are the X-ray diffraction patterns produced by analysis of rock samples in Curiosity's CheMin laboratory. On the left is a sample of windblown dust and sand from a location called Rocknest, while on the right is a sample of freshly drilled bedrock obtained at John Klein. Analysis of the patterns allowed mission scientists to identify important differences between the samples —in particular the presence of clay minerals called phyllosilicates in the John Klein rock (the light blue haze at the bottom of the right-hand image), which could only have formed in a relatively hospitable, wet environment

Future Mars missions

The exploration of Mars by robot spacecraft seems set to continue apace over the coming decades. Some of these missions are already on their way to Mars, or at least in the advanced stages of construction, but others are still on the drawing board.

During the November 2013 launch window, two spacecraft departed Earth on the long journey to Mars—both are orbiters, planned to arrive at their destination in September 2014, but each, if successful, will break new ground in the exploration of the Red Planet.

MAVEN (Mars Atmosphere and Volatile Evolution) is a NASA spacecraft designed, as its name suggests, to study the current thin Martian air in great detail. By profiling its structure and precise composition, scientists hope that MAVEN will help them to understand the processes that have influenced the development of the atmosphere throughout its long history, and still do so today. In particular, the satellite aims to measure the distribution of varying "isotopes"—chemically identical atoms that have different masses and therefore behave in physically different ways—in the Martian atmosphere. These may hold the key to explaining how and when the planet lost the bulk of its atmosphere and water, and precisely where it went.

The Mars Orbiter Mission, meanwhile (also known as Managalyaan), is India's first interplanetary probe. This ambitious step forward for the Indian Space Research Organization (ISRO) is primarily aimed at simply overcoming the challenges of getting to Mars, but the mission also carries instruments to analyze the atmosphere and image the surface.

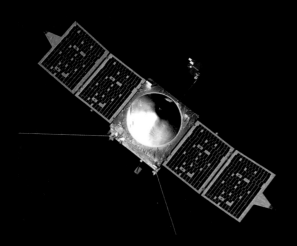

↑ NASA's MAVEN probe, shown here orbiting Mars in an artist's impression, will explore the structure and composition of the Martian atmosphere in detail. Skimming through the upper layers of the atmosphere at altitudes ranging from 93 to 311 miles (150 to 500 kilometers), MAVEN will not just analyze the gases from a distance, but will also collect them for onboard study, helping to answer long-standing questions about the origin and fate of the early Martian atmosphere. The mission is a forerunner for a new generation of Martian explorers with precisely targeted scientific objectives.

↑ The European Space Agency's ExoMars rover is still under development, with prototypes such as the one shown here being tested in the harsh conditions of Chile's Atacama Desert. Smaller than Curiosity but about the same size as NASA's earlier Mars Exploration Rovers, the six-wheeled ExoMars will use solar power for its planned six-month mission. As well as imaging systems, a combined drill and subsurface spectrometer, and ground-penetrating radar, ExoMars will carry the Pasteur Instrument Suite—a set of experiments designed to search specifically for signs of past or present life.

→ The idea of a "weather balloon" probe floating in the atmosphere was suggested for a planned Russian Mars mission as early as the mid-1990s, and seems likely to happen at some point in the future. The balloon would rise into the air during the day, recording atmospheric data, then sink back to the surface at night, allowing instruments suspended beneath it to study the Martian soil wherever it came to rest.

→→ This more speculative design for an aerial Mars probe is ARES (Aerial Regional-scale Environmental Survey of Mars) — a "Mars airplane" mission designed at NASA's Langley Research Center. After entering the atmosphere in a conventional aeroshell, ARES would be released high above the surface and unfold its wings to begin a preprogramed survey pattern during a long glide across up to 380 miles (610 kilometers) of the Martian landscape. Such a mission could provide unique new insights into questions such as the magnetism of Martian rocks, the dynamics of the lower atmosphere, and the role of water vapor.

The next launch window, in the spring of 2016, will see NASA dispatching its InSight lander, and the European Space Agency (ESA), in collaboration with the Russian Federal Space Agency (Roscosmos), launching a two-element mission consisting of the Trace Gas Orbiter spacecraft and a "proof-of-concept" lander called Schiaparelli. InSight will carry a variety of instruments for studying the Martian interior, including a seismometer and a thermal probe, while Schiaparelli will carry advanced meteorological instruments for learning more about the Martian weather and the hazards posed by dust. The Trace Gas Orbiter (TGO), meanwhile, will continue to improve our knowledge of the Martian atmosphere, and may resolve the confusing questions surrounding the possible presence of methane once and for all.

If all goes well, ESA and Roscosmos plan to follow this mission in 2018 with the ExoMars rover—an ambitious six-wheeled vehicle that will likely be targeted at any area where TGO can confirm the existence of methane, and will rely on the earlier satellite to relay its signals to Earth. Instruments aboard the rover are being specifically designed to look for signs of life in these areas. NASA, meanwhile, has plans to launch another Curiosity-style rover in 2020, and beyond that various proposals for sample-return missions, bringing either pristine rock or even scoops of Martian atmosphere back to Earth for analysis.

More speculative plans for extending the long-term exploration of Mars include the idea of sending small "aerobot" probes attached to weather balloons that would carry them above the Martian landscape, perhaps for weeks or even months. Beyond this, it might even be possible to deploy lightweight solar-powered aircraft into the Martian atmosphere. Despite the thin air, a "flying wing" with sufficient surface area and a light enough structure could remain aloft for many months.

Almost inevitably, some of these missions will end in failure, and others may never get off the drawing board. However we can be sure that those probes that do succeed will continue to surprise us with startling new discoveries and deeper insights into the history, geology, and environment of the Red Planet.

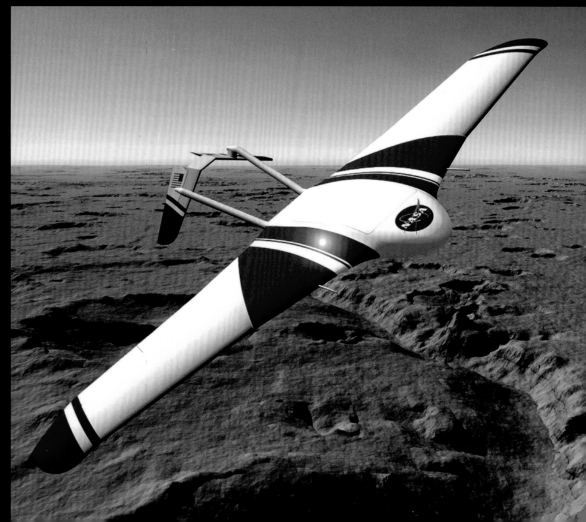

Humans on Mars

Mars is the most welcoming habitat in the solar system beyond Earth, and the logical next step on the human journey into space. Getting there will be a challenge, but getting back will be even more of one—so it may just be easier to stay.

For the rocket enthusiasts of the early 20th century who became the driving forces of the Space Race, Mars was always the ultimate goal for the first wave of human space exploration. Even while NASA was planning the Apollo moonshots of the late 1960s, many of its engineers were talking with some confidence of plans to reach Mars by the 1980s.

The fact that such ideas today seem like wild fantasies is not so much a reflection of the overreaching ambition of those early dreamers, as it is of harsh economic reality and a failure of political willpower. The great Space Race between the United States and the Soviet Union effectively came to an end when Neil Armstrong took that first small step onto the Moon in 1969, and the decades that followed saw manned space flight retrench to Low Earth Orbit (LEO), focusing on more parochial matters while the business of exploring the solar system was left to increasingly sophisticated robot spacecraft.

But the dream of a manned mission to Mars still lingers and it will undoubtedly happen eventually.

And whether the first person to set foot in the dust of the Red Planet will be a US astronaut, a Russian cosmonaut, or a Chinese taikonaut, their mission will ultimately have to overcome the same basic challenges, most of which are dictated by the inexorable physics of planetary orbits. While a manned trip to Mars could feasibly be accomplished in six or seven months—about the same time that space probes currently take—the close approaches required for this short trip only take place every two years or more. The explorers would have to settle in for a long wait on Mars, and the round trip would take over three years in total, even assuming all the technological hurdles could be overcome.

First among these is the sheer difficulty of being able to return at all: sending a manned capsule to Mars is one thing, but supplying it with enough fuel to make a safe landing on the surface, blast off back into orbit, and then return to Earth is quite another. Any "all-in-one" spacecraft attempting to make the trip would be too large to launch directly from Earth, so would almost certainly be assembled in Earth orbit, in which case a more efficient option

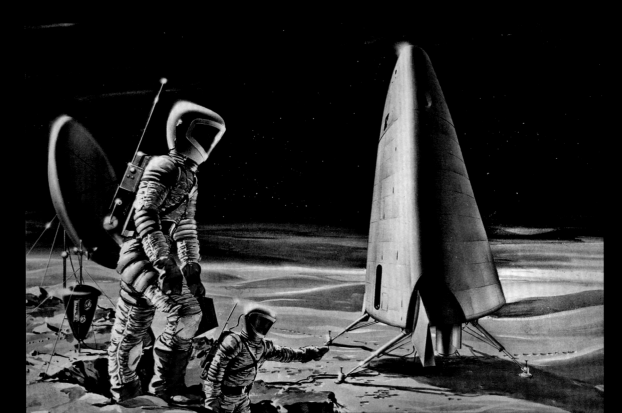

← This artist's impression shows astronauts exploring close to a Mars Excursion Module proposed as part of a detailed NASA study in 1964. At the time, engineers considered the mid-1970s to be a realistic timeframe to achieve a landing on Mars—although there are huge technical difficulties to be overcome, the principal reasons we have not yet reached Mars have been financial and political.
→ An artist's rendering shows NASA's new spacecraft, Orion (at far left) docking with a much larger Mars Transfer Vehicle in Earth orbit. This mission concept, developed in the 2000s as part of NASA's now-defunct Constellation program, would have used a revolutionary "nuclear thermal" propulsion system to take astronauts to Mars within six months around favorable orbital alignments.

might be to send a fully fueled return vehicle on ahead to Mars and ensure that it is safely in orbit before dispatching the human explorers.

One ingenious take on this problem is a mission profile known as "Mars Direct," developed by Robert Zubrin in the 1990s. This envisages sending an Earth Return Vehicle (ERV) not merely to wait in Martian orbit, but to land on the surface and manufacture fuel for the return journey from the raw materials of the Martian soil.

A six-month journey through interplanetary space also presents medical challenges for the people who undertake it. Although the spacecraft could well be spun around its axis in order to generate some basic artificial gravity, it would probably not be practical to produce Earthlike (or even Marslike) gravity continuously, so the crew would have to endure the dangers of muscle weakness and bone wasting experienced by astronauts on the International Space Station. These would be especially risky on arrival at Mars, where the crew would return to an environment with gravity, but

with no means of outside help should one of them suffer an injury. Another significant risk would come from high-energy solar radiation and highly penetrating particles known as cosmic rays. Earth's magnetic field usually shields spacecraft in LEO from these dangers, but a spacecraft on its way to Mars would be unprotected and would almost certainly need to carry appropriate shielding.

In order to minimize the risks involved, the spacefarers might arrive to find most of their equipment waiting for them in a habitat module sent on ahead of the main mission. Given the transport challenges, this Martian base would need to be self-sufficient from the outset, instead of relying on supplies from Earth. Fortunately, recent discoveries have shown that Mars has most of the raw materials needed to support a short-term outpost, including plentiful water, which not only would irrigate plant crops and provide drinking water, but could also be processed to generate oxygen for breathing, and even fuel and electricity. Early Mars explorers might well base themselves near the easy water supplies of an ice-rich

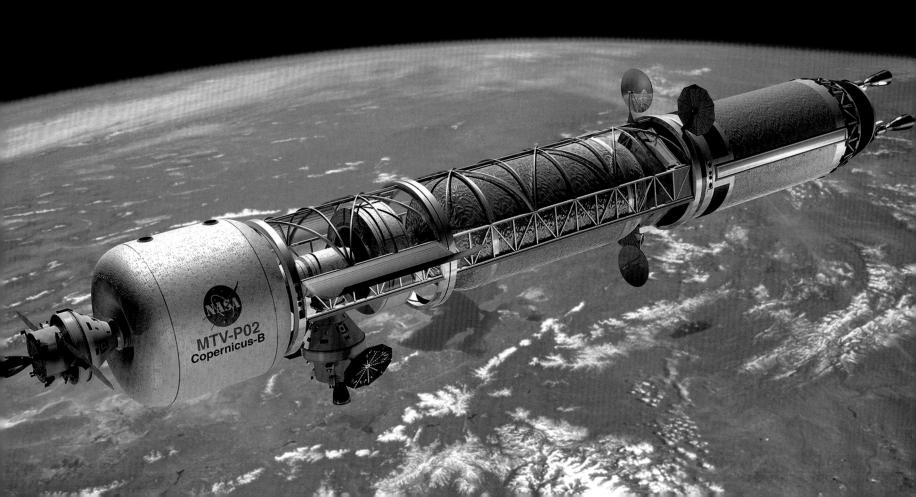

"lobate debris apron" (see page 122) rather than attempting to extract water from the soil itself.

Clearly, exploring Mars is a much bigger undertaking than a short trip to the Moon and back—even preparing for the mission would require multiple equipment launches over several years, and the expedition itself would take years rather than days. On the other hand, fortunately the environment is far more welcoming and the scientific rewards are potentially even more exciting. Once the first expedition has made it safely home it will only be a matter of time before the foundation of the first semipermanent scientific outpost.

In the longer term, could humans one day migrate to Mars in greater numbers? Such ideas are the stuff of science fiction, of course, but no less an authority than the cosmologist Stephen Hawking has argued that they are necessary for the long-term survival of the human race. He predicts that "we will eventually establish self-sustaining colonies on Mars and other bodies in the solar system, but not within the next 100 years."

Wholesale colonization of Mars would almost certainly require the large-scale reengineering of its environment—a procedure known as "terraforming." NASA scientists have already proposed ways in which this could be done. One proposal is to deliberately pump greenhouse gases into the thin air to trigger a runaway global warming effect that would melt the polar caps. This would release carbon dioxide and water to thicken the planet's atmosphere, while planting trees would process carbon dioxide into breathable oxygen. Such a process might take many thousands of years to complete, but it would begin with that first single footstep in the Martian dust.

↑ This artist's rendering accompanied a 2009 NASA study for a "Mars Reference Mission." Recognizing the fact that any mission to Mars will necessarily involve a long stay on the planet's surface, the study outlined exploration vehicles including a pressurized Mars rover and larger habitation modules separate from the landing vehicle itself.
↓ Another artist's rendering shows an Orion spacecraft docking with an orbiting transfer rocket for the long journey back to Earth.

→ Could this be a vision of Mars in the far future? With its atmosphere thickened by centuries of terraforming, water might once again flow on the Martian landscape, perhaps even forming shallow oceans and flooding the mighty Valles Marineris. In this scenario, Mars could become a second Earth, and perhaps a vital lifeboat for our long-term cosmic survival.

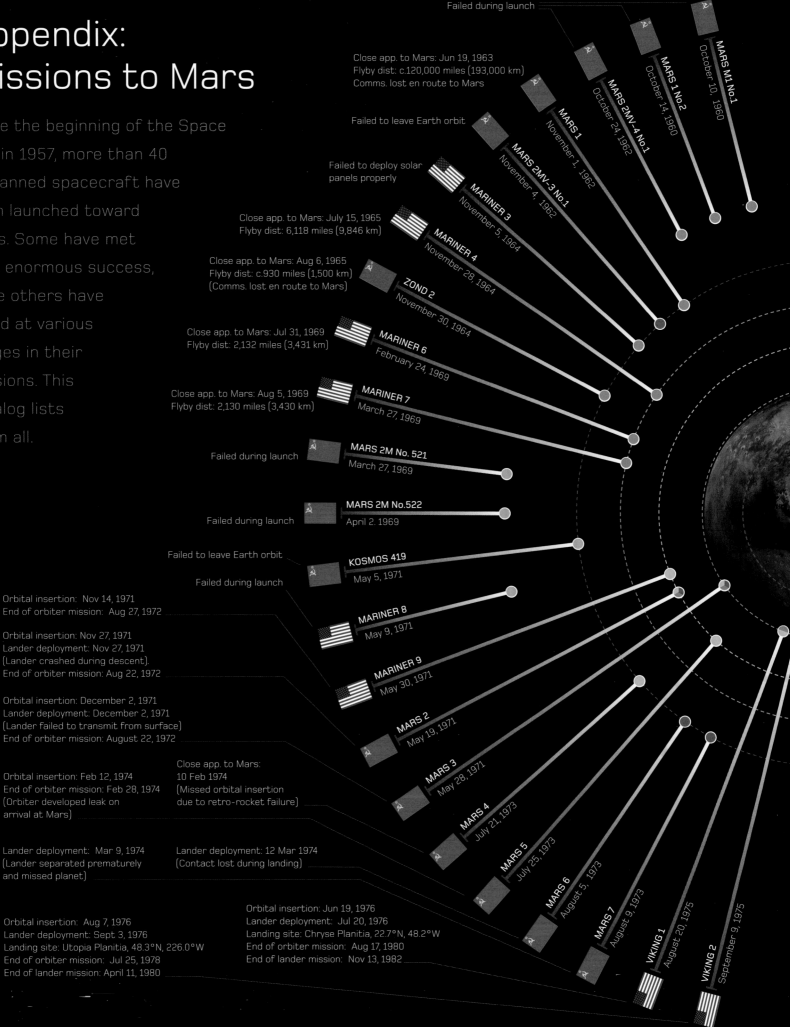

Appendix: Missions to Mars

Since the beginning of the Space Age in 1957, more than 40 unmanned spacecraft have been launched toward Mars. Some have met with enormous success, while others have failed at various stages in their missions. This catalog lists them all.

Failed during launch

MARS M1 No.1
October 10, 1960

MARS 1 No.1
October 14, 1960

MARS 2MV-4 No.1
October 24, 1962

Close app. to Mars: Jun 19, 1963
Flyby dist: c.120,000 miles (193,000 km)
Comms. lost en route to Mars

MARS 1
November 1, 1962

Failed to leave Earth orbit

MARS 2MV-3 No.1
November 4, 1962

Failed to deploy solar panels properly

MARINER 3
November 5, 1964

Close app. to Mars: July 15, 1965
Flyby dist: 6,118 miles (9,846 km)

MARINER 4
November 28, 1964

Close app. to Mars: Aug 6, 1965
Flyby dist: c.930 miles (1,500 km)
(Comms. lost en route to Mars)

ZOND 2
November 30, 1964

Close app. to Mars: Jul 31, 1969
Flyby dist: 2,132 miles (3,431 km)

MARINER 6
February 24, 1969

Close app. to Mars: Aug 5, 1969
Flyby dist: 2,130 miles (3,430 km)

MARINER 7
March 27, 1969

Failed during launch

MARS 2M No. 521
March 27, 1969

Failed during launch

MARS 2M No.522
April 2. 1969

Failed to leave Earth orbit

KOSMOS 419
May 5, 1971

Failed during launch

Orbital insertion: Nov 14, 1971
End of orbiter mission: Aug 27, 1972

MARINER 8
May 9, 1971

Orbital insertion: Nov 27, 1971
Lander deployment: Nov 27, 1971
(Lander crashed during descent).
End of orbiter mission: Aug 22, 1972

MARINER 9
May 30, 1971

Orbital insertion: December 2, 1971
Lander deployment: December 2, 1971
(Lander failed to transmit from surface)
End of orbiter mission: August 22, 1972

MARS 2
May 19, 1971

MARS 3
May 28, 1971

Orbital insertion: Feb 12, 1974
End of orbiter mission: Feb 28, 1974
(Orbiter developed leak on
arrival at Mars)

Close app. to Mars:
10 Feb 1974
(Missed orbital insertion
due to retro-rocket failure)

MARS 4
July 21, 1973

MARS 5
July 25, 1973

Lander deployment: Mar 9, 1974
(Lander separated prematurely
and missed planet)

Lander deployment: 12 Mar 1974
(Contact lost during landing)

MARS 6
August 5, 1973

MARS 7
August 9, 1973

Orbital insertion: Aug 7, 1976
Lander deployment: Sept 3, 1976
Landing site: Utopia Planitia, 48.3°N, 226.0°W
End of orbiter mission: Jul 25, 1978
End of lander mission: April 11, 1980

Orbital insertion: Jun 19, 1976
Lander deployment: Jul 20, 1976
Landing site: Chryse Planitia, 22.7°N, 48.2°W
End of orbiter mission: Aug 17, 1980
End of lander mission: Nov 13, 1982

VIKING 1
August 20, 1975

VIKING 2
September 9, 1975

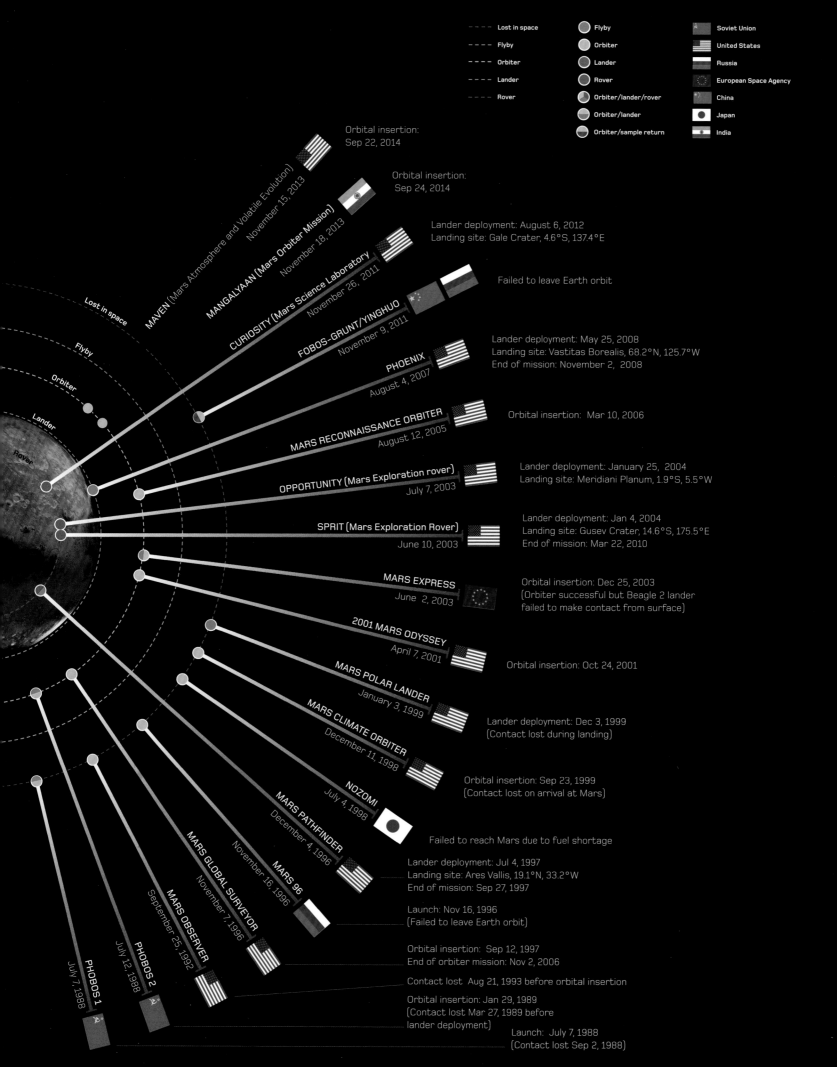

Lost in space

Flyby

Orbiter

Lander

Rover

Flyby

Orbiter

Lander

Rover

Orbiter/lander/rover

Orbiter/lander

Orbiter/sample return

Soviet Union

United States

Russia

European Space Agency

China

Japan

India

Orbital insertion:
Sep 22, 2014

Orbital insertion:
Sep 24, 2014

Lander deployment: August 6, 2012
Landing site: Gale Crater, 4.6°S, 137.4°E

Failed to leave Earth orbit

Lander deployment: May 25, 2008
Landing site: Vastitas Borealis, 68.2°N, 125.7°W
End of mission: November 2, 2008

Orbital insertion: Mar 10, 2006

Lander deployment: January 25, 2004
Landing site: Meridiani Planum, 1.9°S, 5.5°W

Lander deployment: Jan 4, 2004
Landing site: Gusev Crater, 14.6°S, 175.5°E
End of mission: Mar 22, 2010

Orbital insertion: Dec 25, 2003
(Orbiter successful but Beagle 2 lander
failed to make contact from surface)

Orbital insertion: Oct 24, 2001

Lander deployment: Dec 3, 1999
(Contact lost during landing)

Orbital insertion: Sep 23, 1999
(Contact lost on arrival at Mars)

Failed to reach Mars due to fuel shortage

Lander deployment: Jul 4, 1997
Landing site: Ares Vallis, 19.1°N, 33.2°W
End of mission: Sep 27, 1997

Launch: Nov 16, 1996
(Failed to leave Earth orbit)

Orbital insertion: Sep 12, 1997
End of orbiter mission: Nov 2, 2006

Contact lost Aug 21, 1993 before orbital insertion

Orbital insertion: Jan 29, 1989
(Contact lost Mar 27, 1989 before
lander deployment)

Launch: July 7, 1988
(Contact lost Sep 2, 1988)

MAVEN (Mars Atmosphere and Volatile Evolution)
November 15, 2013

MANGALYAAN (Mars Orbiter Mission)
November 18, 2013

CURIOSITY (Mars Science Laboratory)
November 26, 2011

FOBOS-GRUNT/YINGHUO
November 9, 2011

PHOENIX
August 4, 2007

MARS RECONNAISSANCE ORBITER
August 12, 2005

OPPORTUNITY (Mars Exploration rover)
July 7, 2003

SPRIT (Mars Exploration Rover)
June 10, 2003

MARS EXPRESS
June 2, 2003

2001 MARS ODYSSEY
April 7, 2001

MARS POLAR LANDER
January 3, 1999

MARS CLIMATE ORBITER
December 11, 1998

NOZOMI
July 4, 1998

MARS PATHFINDER
December 4, 1996

MARS 96
November 16, 1996

MARS GLOBAL SURVEYOR
November 7, 1996

MARS OBSERVER
September 25, 1992

PHOBOS 2
July 12, 1988

PHOBOS 1
July 7, 1988

Appendix: Observing Mars

Even today, when the bulk of scientific research relies on complex space probes and enormous professional telescopes, Mars remains a hugely popular target for amateur astronomers, with something to offer all levels of observer.

At its brightest and closest to Earth, Mars can outshine every object in the night sky apart from the Moon and Venus, and even when it is noticeably fainter and more distant its blood-red hue makes it more prominent than its brightness alone would suggest, and easily identifiable. Like all the major planets, Mars moves slowly against the "fixed" background stars, completing a circuit of the sky roughly once every 26 months. Its path around the sky sticks close to a line known as the ecliptic, which marks the Sun's own annual track through the stars (in reality, this track is defined by Earth's own orbit around the Sun, but since all the major planets orbit in roughly the same flat plane, they are always found close to the ecliptic). To the naked eye, Mars's most intriguing feature is its prominent "retrograde" motion: while planets generally move from eastward against the background stars, Mars can reverse its path and describe large westward loops around the time of opposition (closest approach to Earth), before resuming its eastward drift. This effect is caused by the faster-moving Earth "undertaking" Mars around opposition, so that the planet appears to move backward from our point of view. In reality, all the planets beyond Earth's orbit show retrograde loops, but Mars's are the largest and longest by far, and the struggle to explain them played a key role in an astronomical revolution (see page 10). Unfortunately, the Red Planet's small size means it presents only a tiny disk to Earth even at its closest. In general, its size ranges between roughly ⅟₆₀ the size of the Full Moon at opposition and ⅟₄₀₀ of a Full Moon near superior conjunction when Mars is on the opposite side of the Sun to Earth. Nevertheless, binoculars will always show Mars as a rock-steady, if tiny, circle of light with none of the unsteady "twinkle" often associated with stars. A small but good-quality telescope, on the other hand, will really bring Mars to life—especially around the favorable oppositions listed in the table below. Refracting (lens-based) telescopes are generally better than reflecting (mirror-based) instruments for detailed planetary observations,

but any decent telescope should be able to show the bright spots of the polar caps and prominent "albedo features" such as the dark triangle of Syrtis Major (see page 150). Colored eyepiece filters are particularly useful for exaggerating the contrast of different Martian features—for example, orange and red enhance the contrast between light and dark regions, while blue and green darken the planet itself while highlighting atmospheric clouds. Perhaps the most advanced project for the serious Mars observer, however, involves imaging the Red Planet. Traditionally, amateur astronomers have tended to record their observations of planets on sketch maps or attempted to photograph the planet using expensive photographic equipment, but a new wave of digital photography has transformed the possibilities for imaging Mars and other planets. Using little more than a modified webcam, a telescope eyepiece adapter, and freely available software, it's now possible to "stack" large numbers of digital images and apply ingenious image processing to extract detail from them. The results can be astonishing, and are often capable of rivaling those achieved with much larger instruments.

← A series of exposures capture the movement of Mars along its retrograde loop around the time of its 2003 opposition. The variations in the planet's size and brightness as it approaches and draws away from Earth can clearly be seen.

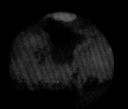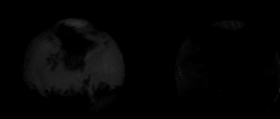

← Simulated views of Mars through orange, red, and blue filters overlaid on an original Hubble Space Telescope image. Orange and red heighten contrast between areas of the planet, while blue highlights atmospheric features.

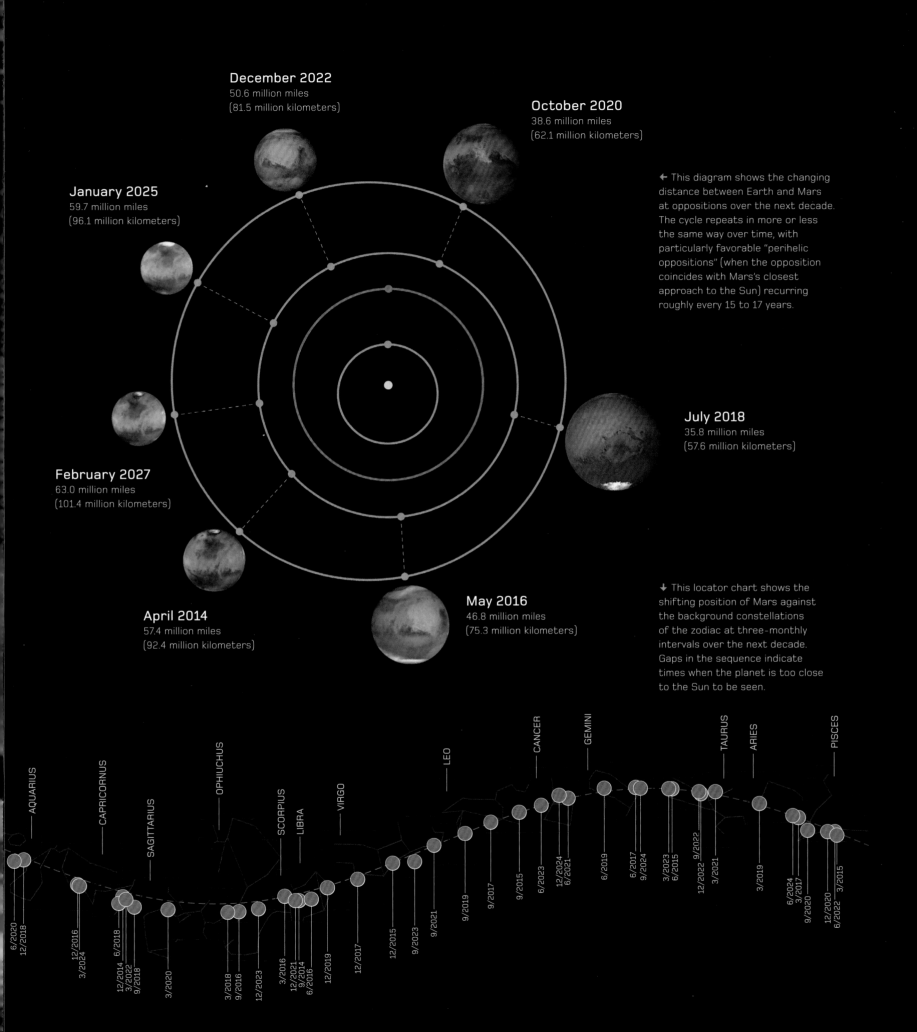

December 2022
50.6 million miles
(81.5 million kilometers)

October 2020
38.6 million miles
(62.1 million kilometers)

January 2025
59.7 million miles
(96.1 million kilometers)

← This diagram shows the changing distance between Earth and Mars at oppositions over the next decade. The cycle repeats in more or less the same way over time, with particularly favorable "perihelic oppositions" (when the opposition coincides with Mars's closest approach to the Sun) recurring roughly every 15 to 17 years.

July 2018
35.8 million miles
(57.6 million kilometers)

February 2027
63.0 million miles
(101.4 million kilometers)

↓ This locator chart shows the shifting position of Mars against the background constellations of the zodiac at three-monthly intervals over the next decade. Gaps in the sequence indicate times when the planet is too close to the Sun to be seen.

April 2014
57.4 million miles
(92.4 million kilometers)

May 2016
46.8 million miles
(75.3 million kilometers)

AQUARIUS
CAPRICORNUS
SAGITTARIUS
OPHIUCHUS
SCORPIUS
LIBRA
VIRGO
LEO
CANCER
GEMINI
TAURUS
ARIES
PISCES

6/2020
12/2018
12/2016
3/2024
12/2014
3/2022
9/2018
3/2020
3/2018
9/2016
12/2023
3/2016
12/2021
9/2014
6/2016
12/2019
12/2017
12/2015
9/2023
9/2021
9/2019
9/2017
9/2015
6/2023
12/2024
6/2021
6/2019
6/2017
9/2024
3/2023
6/2015
12/2022
9/2022
3/2021
3/2019
6/2024
3/2017
9/2020
12/2020
6/2022
3/2015

Aerobraking
The process of altering the shape and speed of a spacecraft's orbit by repeatedly using the air resistance of a planet's upper atmosphere as a brake to slow the vehicle.

Aeroshell
A protective aerodynamic shell around a lander space probe, designed to protect the probe itself from heat generated during entry into a planet's atmosphere, and to slow the vehicle's descent (usually in conjunction with a parachute).

Albedo
An astronomical term measuring the amount of sunlight reflected by planet's surface or atmosphere —planets with high albedos tend to appear brighter.

Amazonian period
The fourth and most recent phase of Martian history, beginning around 3.0 billion years ago and lasting to the present day. During the Amazonian, the planet's surface conditions have been broadly similar to those of the present day, with a thin atmosphere, cold dry surface, and low levels of impact cratering. However, occasional catastrophic floods continued in the early part of this period, and volcanic activity continued until very recently (if not up to the present).

Asteroid
Any small, rocky body following an independent orbit around the Sun. Asteroids range in size from less than 3 feet (1 meter) to several hundred miles across, and are mostly confined to the main Asteroid Belt between Mars and Jupiter.

Atmosphere
A layer of gases surrounding a planet, held in place by its gravity.

Backshell
The rear half of an aeroshell. In modern Mars landers, the backshell often carries a braking system, such as parachutes or retrorockets, and lowers the lander vehicle itself toward the ground on a long tether.

Chaos terrain
A type of Martian landscape characterized by a jumble of mesas, hummocks, and collapsed areas of terrain, thought to be created by catastrophic subsidence as groundwater is drained from beneath the surface.

Cone volcano
A volcano in which layers of ash build up in a conical shape surrounding a single volcanic vent that erupts repeatedly.

Coriolis effect
A "virtual force" that affects motion across a planet's surface, caused by the fact that the planet provides a rotating "frame of reference." On Mars and Earth, the Coriolis effect causes objects (including air masses and weather systems) to be deflected clockwise from their direction of travel in the northern hemisphere, and counterclockwise in the southern hemisphere.

Cosmic rays
High-speed particles emitted from a variety of astronomical objects including our Sun, which have the ability to penetrate spacecraft hulls and damage both electronic systems and living tissue.

Crustal dichotomy
The broad division between two types of terrain on Mars—rugged, heavily cratered highlands in the planet's southern hemisphere, and smooth lowland plains in the north.

Datum
The average Martian surface elevation, used as a baseline for height measurements on Mars in the same way as sea level is used on Earth.

Ejecta
Material thrown out across a planet's surface by a violent event (the term usually refers to the debris flung out during meteorite impacts, but ash and hot rocks produced by volcanic eruptions are also sometimes termed ejecta).

Extremophile
Any organism that thrives in conditions that would seem intolerable to most life on Earth. Extremophiles can tolerate extremes of hot or cold temperature, acid or alkali environments, other normally toxic substances, or even prolonged exposure to vacuum.

Fossa
A long, narrow trough on the surface of an extraterrestrial body. Fossae on Mars are often found in parallel groups, and are often linked to stresses in the crust caused by the weight of nearby volcanoes.

Fretted terrain
A type of terrain found in places along the boundary of the Martian crustal dichotomy, where narrow straight valleys merge into chaos terrain. Fretted terrain is thought to be formed by the action of glaciers.

Graben
A sunken area of planetary surface bounded on either side by parallel faults, where one area of the crust has slipped downward in relation to its neighbors. Grabens usually form in places where a planet's crust is being pulled apart.

Greenhouse gases
Any of a variety of gases that tend to trap solar heat near a planet's surface and prevent it from escaping into space, increasing its average temperature. Familiar greenhouse gases include carbon dioxide and methane.

Ground-penetrating radar
A remote-sensing instrument that fires radio waves that penetrate some way into a planet's surface before reflecting back, allowing the upper layers of the planet's crust to be mapped in some detail.

Gully
A Martian landscape feature linked to the presence of liquids on the planet's surface. Gullies usually originate in a scoop-shaped alcove on a steep slope, from which a single channel runs downhill to eventually discharge in a fan of debris. While some gullies can be linked to the behavior of carbon dioxide frosts, others are almost certainly the result of running water.

Hematite
An iron oxide mineral that is only known to form underwater, and which therefore provides some of the strongest evidence for prolonged wet periods in the Martian past.

Hesperian period
The third great stage of Martian history, characterized by the disappearance of exposed surface water and a thinning of the atmosphere coupled with widespread volcanic eruptions and frequent catastrophic floods from escaping groundwater. The Hesperian is thought to have lasted from around 3.7 billion to 3.0 billion years ago, though dating Martian geological history is highly uncertain.

Hot spot
A high-temperature region at the top of a planet's mantle where it meets the crust. Hot spots are thought to be caused by plumes of hot material rising through the mantle from close to the core—their heating effect can trigger volcanic activity in the overlying crust.

Hydrothermal
Associated with natural hot water such as that heated by volcanism. Hot springs and geysers are both forms of hydrothermal activity that create environments hospitable for certain forms of life.

Igneous rock
A type of rock formed from the cooling and solidification of molten volcanic magma or lava.

Interior layered deposit

A layered sedimentary rock built up on a valley or crater floor by the deposition of sediments that may have been blown there by the wind or washed there at a time when the area was underwater. Often, such deposits have subsequently been eroded away by later geological processes.

Inverted relief

A description of any landscape feature whose usual topography is inverted so that normally elevated areas are sunken, while normally lower areas are raised above their surroundings. Inverted features on Mars include riverbeds and craters, both of which are thought to result from the formation of sedimentary rock that is highly resistant to later erosion.

Kuiper Belt

A region of the outer solar system containing countless small, ice-dominated worlds such as Pluto. Today, Kuiper Belt Objects are mostly confined beyond the orbit of Neptune, but in the early days of the solar system they formed much closer to the Sun.

Laser altimeter

A remote-sensing instrument that fires a laser at a planetary surface and measures the time taken to receive the reflection, building up a detailed model of the distance (and therefore topography) of the surface below.

Late Heavy Bombardment

A sudden peak in the rate and size of impacts on planetary surfaces throughout the inner solar system, thought to have occurred around 4 billion years ago, and now linked to the migration of the outer giant planets into the Kuiper Belt.

Lobate debris apron

A lobe-shaped feature found at the foot of steep slopes in mid-latitude Martian environments. Consisting of ice covered in a thin layer of dust and rock, debris aprons are a form of Martian glacier.

Low Earth Orbit

An orbit around just a few hundred miles above the surface of Earth, within the protection of our planet's magnetic field, where most satellites and manned spacecraft fly.

Mensa

A term (from the Latin for "table") applied to mesalike features on Mars and other solar system bodies.

Mesa

A landscape feature with a flat top and steep cliffs on its sides—left behind as an isolated plateau when its surroundings have eroded away or otherwise collapsed.

Meteorite

A rock from space that survives entry into a planetary atmosphere (where it may briefly appear as a meteor or shooting star) and lands on a planet's surface. Most meteorites are fragments of small asteroids, but a few originate on the surface of other large planetary bodies.

Methanogen

A type of Earth microbe that can process carbon dioxide and hydrogen in the absence of oxygen to produce methane—a completely different metabolic pathway from most Earth organisms. Methanogens are often extremophiles, capable of thriving in harsh environments.

Noachian period

The second broad phase of Martian history—a period of heavy bombardment from space, and probably abundant surface water. The Noachian is thought to correspond with the Late Heavy Bombardment found on the Moon, and to have lasted between 4.1 billion and 3.7 billion years ago.

Opposition

An alignment of planetary orbits in which the planet is direct opposite the Sun in Earth's skies. At this time, a planet such as Mars is at its closest to Earth.

Outflow channel

A broad valley, often with deeper channels within it and isolated teardrop-shaped islets, which marks a sudden and catastrophic release of water onto the Martian surface.

Patera

From an ancient type of bowl, a patera is a type of volcano on Mars that consists of a collapsed central caldera with only a very low surrounding volcanic shield (if any).

Polar layered deposit

A type of layered terrain underlying both Martian poles, formed by the annual deposition of layers of dust and ice to ultimately form a plateau several miles high.

Pre-Noachian period

The oldest period of Martian history, during which the cooling planet had abundant surface water and was subject to steady but declining meteorite bombardment. The Pre-Noachian lasted from the formation of Mars around 4.5 billion years ago, to the beginning of the Noachian about 4.1 billion years ago.

Recurrent slope lineae

Dark streaks that spread out across parts of the Martian landscape in the spring and summer of each year before fading away. Despite appearances, they are not wet themselves, but are thought to be created by liquid water flowing just below the surface.

Retrorocket

A rocket that points in the opposite direction to a spacecraft's motion, and which therefore acts as a brake when it is fired.

Sedimentary rock

A type of rock formed when fine-grained sediments are allowed to build up in layers over time (often underwater or in a natural depression) and subsequently compressed under their own weight.

Shield volcano

A large volcano in which lava from an underlying magma chamber erupts over long periods through multiple vents to form a broad, shallow "shield." Once the magma chamber is emptied, the top of the shield often collapses to form a roughly circular caldera.

Sol

A Martian day, equivalent to 24 hours 37 minutes 22.7 seconds.

Solar radiation

Electromagnetic radiation from the Sun, including not only visible light, but also infrared (heat) radiation, ultraviolet rays, and X-rays.

Spectrometer

A device that measures the specific wavelengths and energies of light that an object emits, reflects, or absorbs—behavior that is linked to the chemical composition of the substances within it.

Sublimation

The process in which a substance changes phase directly from a solid into a gas without passing through an intermediate liquid phase. Frozen carbon dioxide ("dry ice") normally sublimes in this way on both Mars and Earth.

Superior conjunction

An alignment of planetary orbits in which a planet is on the far side of the Sun from Earth, and therefore at its greatest distance from us.

Terraforming

The hypothetical process of transforming a planet's surface environment to mirror Earth's, in order to allow easier human colonization.

Tholus

A term used for classifying small dome-shaped mountains and hills on Mars, often with a volcanic origin.